Interventional Ultrasound

Edited by

HANS HENRIK HOLM

JØRGEN KVIST KRISTENSEN

SPRINGER-VERLAG BERLIN HEIDELBERG GMBH

Interventional Ultrasound
1st edition

Copyright © 1985 Springer-Verlag Berlin Heidelberg
Ursprünglich erschienen bei Munksgaard, Copenhagen 1985
Softcover reprint of the hardcover 1st edition 1985

Cover by Lars Thorsen
Typesetting: P. J. Schmidt, Vojens
Reproduction: Odense Reproduktion

ISBN 978-87-16-09776-7   ISBN 978-3-662-25530-8 (eBook)
DOI 10.1007/978-3-662-25530-8

Interventional Ultrasound

Contributors

Peter Alken M.D.
Professor of Urology
Urologische Klinik und Poliklinik
der Johannes Gutenberg-Universität Mainz
Langenbeckstr. 1, 6500 Mainz
Germany

Yasutsugu Bandai, M.D.
Second Department of Surgery
University of Tokyo
7-3-1 Hongo Bunkyoku, Tokyo 113
Japan

Jens Bang, M.D.
Chief Obstetrician
Director of Dept. for Diagnostic
Ultrasound YU 4023
Rigshospitalet, University of Copenhagen
Blegdamsvej 9, 2100 Copenhagen
Denmark

Branco Breyer, Ph.D.
Gynecological Cancer Center
University of Zagreb, Zagreb
Yugoslavia

Ivo Cikes, M.D., Ph.D.
Associate Professor of Medicine
Head of Echocardiographic Laboratory
Institute of Cardiovascular Diseases
University Hospital Rebro, Zagreb
Yugoslavia

William Clewell, M.D.
Associate Professor of OB/GYN
University of Colorado Health
Sciences Center
Denver, CO 80262
USA

Feder Custovic, M.D., Ph.D.
Professor of Medicine
Director of Institute of Cardiovascular
Diseases
University Hospital Rebro, Zagreb
Yugoslavia

Alexander Ernst, M.D., Ph.D.
Institute of Cardiovascular Diseases
University Hospital Rebro, Zagreb
Yugoslavia

Sven Grønvall, M.D.
Department of Radiology
Herlev Hospital, Univerity of Copenhagen
2730 Herlev, Copenhagen
Denmark

Bo Hainau, M.D.
Consultant,
Department of Pathological Anatomy
Herlev Hospital, University of Copenhagen
2730 Herlev, Copenhagen
Denmark

Søren Hancke, M.D., Ph.D.
Director of Ultrasound Laboratory
Gentofte Hospital,
University of Copenhagen
2900 Hellerup, Copenhagen
Denmark

Hans Erik Hansen, M.D.
Department of Medicine C
Municipal Hospital of Aarhus, 8000 Aarhus
Denmark

Hiroshi Hasegawa, M.D.
Department of Hepatic Surgery
National Cancer Center Hospital
5-1-1 Tsukiji Chuo-ku, Tokyo 104
Japan

Aksel Haubek, M.D.
Director of Division of CT and
Ultrasound,
Department of Radiology
Municipal Hospital of Aarhus
8000 Aarhus
Denmark

Anders Hemmingsson, M.D.
Department of Diagnostic Radiology
Uppsala University, Uppsala
Sweden

Hans Henrik Holm, M.D., Ph.D.
Director of Department of Ultrasound
Chief Surgeon, Department of Urology H
Herlev Hospital, University of Copenhagen
2730 Herlev, Copenhagen
Denmark

Grete Krag Jacobsen, M.D.
Department of Pathological Anatomy
Hvidovre Hospital,
University of Copenhagen
2650 Hvidovre, Copenhagen
Denmark

Flemming Jensen, M.D.
Consultant, Diagnostic Ultrasound
Rigshospitalet, Finsen Institute,
University of Copenhagen
Blegdamsvej 9, 2100 Copenhagen
Denmark

Niels Juul, M.D.
Senior Resident, Department of Ultrasound
Herlev Hospital, University of Copenhagen
2730 Herlev, Copenhagen
Denmark

Finn Koch, M.D., Ph.D.
Professor,
Department of Pathological Anatomy
Herlev Hospital,
University of Copenhagen
2730 Herlev, Copenhagen
Denmark

Jørgen Kvist Kristensen, M.D., Ph.D.
Chief Surgeon, Department of Urology D
Rigshospitalet, University of Copenhagen
Blegdamsvej 9, 2100 Copenhagen
Denmark

Svend Larsen, M.D., Ph.D.
Chief Pathologist,
Department of Pathological Anatomy
Herlev Hospital, University of Copenhagen
2730 Herlev, Copenhagen
Denmark

Torben Larsen, M.D.
Research Fellow, Department of Ultrasound
Herlev Hospital, University of Copenhagen
2730 Herlev, Copenhagen
Denmark

Suzan Lenz, M.D.
Department of Diagnostic Ultrasound
Rigshospitalet, University of Copenhagen
Blegdamsvej 9, 2100 Copenhagen
Denmark

Masatoshi Makuuchi, M.D.
Department of Hepatic Surgery
National Cancer Center Hospital
5-1-1 Tsukiji Chuo-ku, Tokyo 104
Japan

David Manchester, M.D.
Associate Professor of Pediatrics
University of Colorado Health
Sciences Center
Denver, CO 80262
USA

Michael L. Manco-Johnson, M.D.
Associate Professor of Radiology &
Medicine
Director, Division of Diagnostic Ultrasound
University of Colorado Health
Sciences Center
Denver, CO 80262
USA

Jan Fog Pedersen, M.D.
Director of Ultrasound Laboratory
Glostrup Hospital,
University of Copenhagen
2600 Glostrup, Copenhagen
Denmark

Dolores Pretorius, M.D.
Assistant Professor of Radiology
University of Colorado Health
Sciences Center
Denver, CO 80262
USA

Munemasa Ryu, M.D.
Second Department of Surgery
Chiba University School of Medicine
1-8-1 Inohana-cho, Chiba City 280
Japan

Maxwell Sehested, M.D.
Senior Resident,
Department of Pathological Anatomy
Herlev Hospital, University of Copenhagen
2730 Herlev, Copenhagen
Denmark

Edward H. Smith, M.D.
Professor and Chairman
University of Massachusetts
Medical School
55 Lake Avenue North
Worcester, Massachusetts 01605
USA

Kenichi Takayasu M.D.
Department of Diagnostic Radiology
National Cancer Center Hospital
5-1-1 Tsukiji, Chuo-ku, Tokyo 104
Japan

Søren Torp-Pedersen, M.D.
Research Fellow, Department of Ultrasound
Herlev Hospital, University of Copenhagen
2730 Herlev, Copenhagen
Denmark

Mogens Vyberg, M.D.
Senior Resident,
Department of Pathological Anatomy
Hvidovre Hospital,
University of Copenhagen
2650 Hvidovre, Copenhagen
Denmark

Susumu Yamazaki, M.D.
Department of Hepatic Surgery,
National Cancer Center Hospital
5-1-1 Tsukiji Chuo-ku, Tokyo 104
Japan

Preface

Modern sonography makes possible the detection of small and subtle changes in the normal echo pattern. These may represent significant pathological changes which can not always be fully revealed by the echo pattern alone. There is, therefore, an increasing need for the supplement of the ultrasonically guided percutaneous puncture, which can be performed with great accuracy and with virtually no risk. Also, ultrasonically guided puncture has proven invaluable for a wide variety of therapeutic purposes.

The first percutaneous puncture guided by ultrasonic scanning using a specifically designed transducer was performed in 1969 at the ultrasonic laboratory in Gentofte, now Herlev, Hospital, Copenhagen. The idea was based on a puncture transducer described and used by Kratochwill for puncture under the guidance of the ultrasonic A-presentation technique.

The development in the field formed the basis of the First International Conference on Ultrasonically Guided Puncture at Herlev Hospital in 1978 sponsored by The Danish Society of Diagnostic Ultrasound. The knowledge and experience of the speakers at that conference was compiled in the book "Ultrasonically Guided Puncture Technique" published in 1980. Since then the Society has sponsored a conference at the same place in 1980 and 1983, the latest conference being termed "Third International Conference on Interventional Ultrasound". This conference demonstrated to us the impressive expansion of the topic in various ways: an expansion of experience with the well known applications of the technique, giving a better background for stating indications, and discussing possible problems and risks; an expansion of fields of application based on both technical and clinical pioneer work; an enormous expansion in the number of performers and users of the technique.

This has for us suggested the need for an up-dated book on the subject and we have, therefore, asked a number of speakers to review their specific area of interest and experience into separate chapters. It is the aim of the book to thoroughly describe techniques, present results, in special to put the results into a clinical context, and also to point to some possible future application. It has been our intention that the book be equally useful to those who perform interventional ultrasound as to those who send their patients to have it done.

We would like to acknowledge the excellent collaboration with all the contributors, which greatly facilitated the editing of the book.

Copenhagen 1984

H. H. Holm
J. Kvist Kristensen

Contents

1. **Introduction to interventional ultrasound**   13

2. **Procedure and principles in ultrasonically guided puncture**   16
Background and development   16
Prerequisites, equipment and precautions   17
Commonly used needles   18
Safety precautions   20
Performance of fine needle aspiration biopsy   21

3. **Handling of aspirated material**   23
Solid lesion   23
Cystic lesion   23

4. **General principles of fine needle aspiration cytology**   25
Preparation of the smear   25
Fixation and staining of the smear   25
Cytologic evaluation of the smear   26

5. **Electron microscopy of ultrasound guided fine needle biopsies**   29
Method   29
Results   30
Indications   31

6. **Fine needle histological sampling**   35
Histological sampling with a 23 gauge modified Menghini needle   35
Histological processing   37
Experience with microcore biopsy   37
Retrieval rate in cytology and histology   38
Microscopic evaluation of malignant tumors   40
Microscopic evaluation of benign lesions   40
Special staining procedures   40
Time consumption   40
Advantages and limitations of Surecut biopsies   40

7. **Puncture of focal liver lesions**   43
Technique   43
Cystic focal lesions   44
Solid focal lesions   48
Results   49

8. **Ultrasonically guided cholangiography and bile drainage**   54
Ultrasonically guided percutaneous transhepatic cholangiography (UG-PTC)   54
Ultrasonically guided percutaneous transhepatic bile drainage (UG-PTBD)   56
Ultrasonically guided percutaneous transhepatic gall bladder drainage (UG-PTGBD)   61
Change in indication of PTC   63

9. **Intraoperative puncture of the liver guided by ultrasound**   65

Development of intraoperative ultrasonic probes 65
Biopsy of mass lesions in the liver 66
Puncture of the bile duct 66
Hepatectomy and ultrasonically guided puncture 68

10. Ultrasonically guided percutaneous nephropyelostomy 72
Method 72
Material 74
Complications 75

11. Percutaneous nephrolithotomy 77
Percutaneous renal anatomy and access 77
Patient selection and preparation 77
Anesthesia 79
One-step or x-step procedure 79
Puncture 79
Dilation 79
Intrarenal instrumentation 80
Stone removal 80
Postprocedural care 81
Secondary procedures 82
The residual stone rate 82
Complications 82

12. Puncture of renal mass lesions 84
Solid renal masses 84
Cystic renal masses 87

13. Ultrasonically guided renal biopsy 91
Indications 91
Contraindications 92
Technique 92
Results 92

14. Ultrasonically guided prostatic biopsy 94
Biopsy technique and experience 95
Biopsy needles 97

15. Ultrasonically guided puncture of solid pancreatic mass lesions 100
Solid mass lesions 101
Method 101

Material and results 101
Cystic mass lesions 101
Material and results 101
Discussion 103

16. Ultrasonically guided percutaneous pancreatography 106
Instruments and procedure 106
Material 107
Results 107
Illustrative cases 108

17. Puncture of gynecological masses 113

18. Ultrasonically guided aspiration of human oocytes 117
Pretreatment 117
Equipment 118
Aspiration procedure and techniques 118

19. Intrauterine needle diagnosis 122
Indications 122
Method 123
Results 124

20. Amniocentesis in late pregnancy 129
Indications 129
Technique 130
Results and complications 130

21. Fetal therapy 132
Prerequisites for fetal therapy 132
Erythroblastosis fetalis 133
Fetal tachycardia with congestive heart failure 136
Urinary tract obstruction 136
Fetal hydrocephalus 138

22. Puncture of retroperitoneal mass lesions 143
Material and results 146

23. Puncture of gastrointestinal mass lesions 148
Material and results 150

24. Diagnostic and therapeutic puncture of intra-abdominal fluid collections   154
Diagnostic puncture   155
Therapeutic puncture   156

25. Interventional echocardiography   160
Pericardiocentesis guided by echocardiography   160
Percutaneous pericardial biopsy and fenestration   164
Cardiac catheterization guided by ultrasound   164
Ultrasonically guided introduction of pacing and electrophysiologic electrodes   167

26. Fine-needle aspiration biopsy: Are there any risks?   169

Experimental evidence of tumor spread after fine-needle aspiration biopsy – literature review   169
Clinical evidence of hazards – literature review   170
Needle tract seeding – literature review   171
Fatalities – literature review   171
Questionnaire   172
Results – Complications listed in tables 26-1–26-4   172
Discussion   176
Conclusion   176

27. Interventional ultrasound in cancer therapy   178
Radioactive sources   178
Chemical substances   182
Heat   183

CHAPTER 1

Introduction to interventional ultrasound

H. H. Holm & J. Kvist Kristensen

A few years ago a patient with weight-loss, uncharacteristic upper abdominal sensations, slight obstipation and elevated sedimentation rate would probably have spent a couple of weeks undergoing quite a number of laboratory tests and conventional radiological procedures such as barium enema, X-ray of the stomach, a cholecystogram and probably also an IVP, without result. Not infrequently an exploratory laparotomy was the final diagnostic procedure. Although quite efficient it was a strain to the patient and, because of the extended hospital stay, expensive for the tax-payer.

Today the above and numerous other categories of patient will hopefully be sent directly to an abdominal ultrasound examination. The equipment has been through a tremendous development since its early days and now in most cases an ultrasound scanning very gently provides valuable information.

However, although ultrasound scanning can detect pathological lesions, not least in the abdomen, with a high degree of certainty it cannot, for instance, always differentiate between a malignant and a benign lesion. Recently there has been increasing interest in so-called "tissue characterization" with ultra-sound for possible use in such situations.

However, when considering the difficulties frequently encountered with conventional microscopy it does not seem likely that tissue characterization with ultrasound or with any other imaging modality will make it possible, with clinically acceptable accuracy, e.g. routinely to distinguish between a benign and a malignant intraabdominal lesion.

Thus, also for the future, a need exists for a precise, invasive procedure which, virtually without risk, can provide minute tissue samples for cytological, histological or bacteriological examination.

Ultrasound has proved to be superior as a puncture guide to any other imaging modality and an ultrasonically guided biopsy should be considered an integral part of the ultrasound study when additional information is needed.

An ultrasound examination may, e.g. in the above case, demonstrate a solid mass lesion in the region of the pancreas. In the same setting an ultrasonically guided fine needle aspiration biopsy or a histological microcore biopsy is performed (Fig. 1-1). Microscopy of the material obtained from the pancreatic mass may prove the lesion to be cancer. The diagnostic work-up is thereby com-

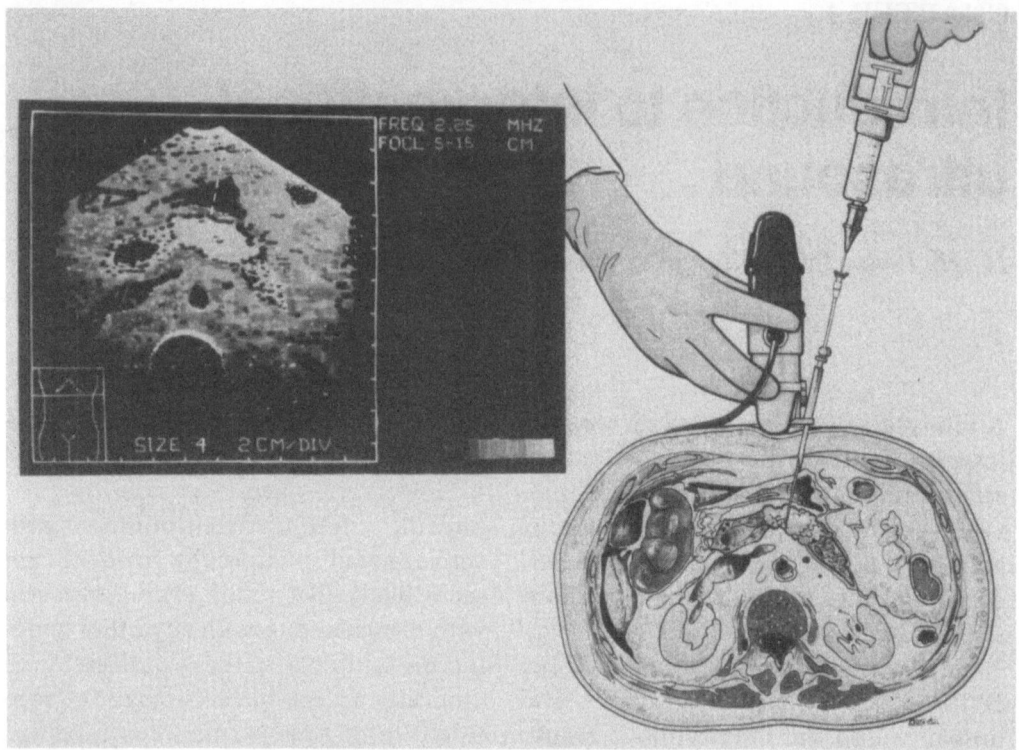

Fig. 1-1. Principle of ultrasonically guided puncture
The scanner is moved and tilted until the image of the target (in this case a pancreatic cancer) is transsected by the electronic punctureline on the monitor. The needle is then inserted through the puncture attachment. In most cases the echo from the needle tip is visible when it proceeds towards the target.

pleted in one setting and a considerable amount of time, incovenience and resources are saved.

The disadvantages of ultrasound as a puncture guide are related to its inability to penetrate gas and bone. Therefore ultrasonic scanning is useless for puncture of lung lesions and in a few cases of abdominal lesions when the patient has severe gaseous distension. It is also useless for bone biopsies.

As shown in the index, ultrasonically guided punctures of a multitude of targets are included in this book, such as the brain, thyroid, heart, all abdominal organs, amniotic fluid, various fetal parts, etc.

A few additional targets could very well have been included such as puncture of non-palpable axillary glands in breast cancer staging, puncture of iliac lymph nodes in staging of prostatic and bladder cancer or puncture of suspected parathyroid adenomas with determination of parathyroid hormone.

Ultrasonically guided puncture may in principle be performed for diagnostic purposes in which situation material may be aspirated for cytological or bacteriological studies or deposited in the form of X-ray contrast in various tubular structures.

Punctures may also be performed for therapeutic reasons, allowing removal of material (e.g. pus) or application of material (e.g. I^{125} seeds and various chemical substances) in cancers. The possibilities are therefore innumerable and every year new applications of the percutaneous puncture technique are found.

As puncture guidance, dynamic ultrasound scanning has many advantages. The method is convenient. The puncture is performed through a small handheld sterile transducer while the "internal part" of the procedure is viewed simultaneously on a TV monitor.

The procedure is rapid. It means, e.g., that during one aspiration biopsy the needle will only stay inside the body for approximately 5 seconds. A whole biopsy procedure, including 5–6 needle passes from various parts of the tumor, smearing etc., usually takes just a few minutes.

A dynamic image is provided. This is especially important when the puncture target is small and moving, such as a renal or a hepatic lesion during respiration.

The needle tip is monitored during its insertion and not, as with CT, checked after its insertion. It means that any inaccuracy in the needle placement is recognized and corrected immediately.

All puncture directions are possible. This is especially important when puncturing upper pole renal lesions and lesions located in the dome of the liver. Here, a cephalad, oblique needle path is mandatory to avoid transpleural puncture with subsequent pneumothorax.

The procedure is completely independent of organ function. This of course is important when a percutaneous nephrostomy is carried out in a patient with postrenal uremia. No ionizing radiation is used, which is an advantage for patient and personnel – not least in obstetrics. The equipment is mobile which means that scans and punctures can be performed in the intensive care unit if indicated. The patient can stay in his bed during the procedure, an obvious advantage over alternative procedures with patients in poor physical condition or in respirators. The technique is relatively inexpensive, which is of particular importance today where cost-benefit factors are under scrutiny.

The procedure has, in our hands, proved to be without risk when fine (0.6 mm) needles are used and with very few complications when 1.2 mm needles (18 gauge) are used.

The only disadvantages of ultrasonically guided puncture are first that the needle tip echo on some occasions can be difficult to identify, probably because the needle bends and leaves the image plane.

Second, that the application of the method is limited by the impermeability of ultrasound through bone and air.

15

CHAPTER 2

Procedure and principles in ultrasonically guided puncture

Flemming Jensen

Background and development

Modern ultrasonic imaging can disclose subtle differences in soft tissue texture and thereby enable the examiner to suggest pathology, but the need for a more specific diagnosis – especially proof or exclusion of malignancy – often arises.

Despite intensive research in tissue characterization by means of image analysis – also with other methods such as CT- and NMR-scanning – it still seems necessary with an interventional method to obtain an exact, specific diagnosis such as a histological one.

This should preferably *not* require an exploratory surgical procedure; whenever possible ultrasonic imaging plus guided puncture should be chosen.

It is the aim in ultrasonically guided puncture to place the tip of an appropriate needle safely and accurately in the suspect lesion or organ, vessel or duct. In its simplest form the puncture is performed free hand after determination of the site and size of the target on the basis of ultrasonic imaging.

In most cases, however, where small, deeply situated lesions are encountered, this method will not prove sufficiently accurate. Thus more precise needle guidance is mandatory.

In the earlier days of static B-mode scanning, a transducer with a central canal was mounted on the scanning arm to steer the needle into a target with a frozen image as guidance and with some support from the "live" A-mode screen. This method, which proved very useful through the 1970's, was developed in Copenhagen by H. H. Holm and coworkers after the development of the first A-mode puncture transducer, presented in Vienna in 1969 by Kratochwill. Simultaneous real-time visualization of the target, the surrounding tissues, part of the needle and its track has become possible with dynamic scanners of different types.

The adjunct of needle steering devices significantly reduces the number of puncture attempts and in most cases the puncture route can be predetermined precisely, so that readjustment of the puncture direction is unnecessary, unless punture of different sites in the target is directly intended.

The practical performance of such punctures was first described by J. F. Pedersen and by M. Saitoh and coworkers in 1977.

16

Prerequisites, equipment and precautions

Given coincidence of the central part of the ultrasonic beam and the needle track, needles of various sizes can be accurately guided to the target from nearly any desired direction without any radiation hazard to operator or patient.

As always in ultrasonic imaging a prerequisite is that the target is visualizable, i.e. not entirely covered by bony or gas-containing structures.

Nowadays most punctures are performed with the aid of high resolution dynamic scanners, which yield a precise monitoring of the puncture procedure.

Mechanical or electronic sector scanners may be used, mounted with a needle steering device, which from one edge of the transducer allows oblique needle steering inside the image plane into the fan-shaped image field. This oblique route is often approx. 20 degrees relative to the vertical centerline of the image field, but in some commercially available models this angle is adjustable (Fig. 2-1).

The ability to predict the needle route depends upon a marker line on the image screen. This is often an electronically generated dotted line which should secure accuracy (Fig. 2-2).

The sector scanners are especially useful when only a limited scan contact area is accessible, but in many instances linear array electronic scanners are sufficient. The needle can be introduced free hand along the array close to the transducer housing or more obliquely from one end of the array. But for precise puncturing of tiny targets, needle steering devices are preferable. These may be built in as slots in the transduceres

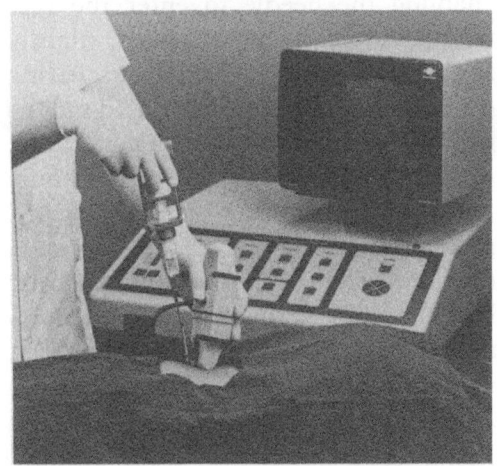

Fig. 2-1. Puncture with mechanical sector scanner
One hand holds the transducer in proper position, while the other moves the needle inside the steering device and effects suction.

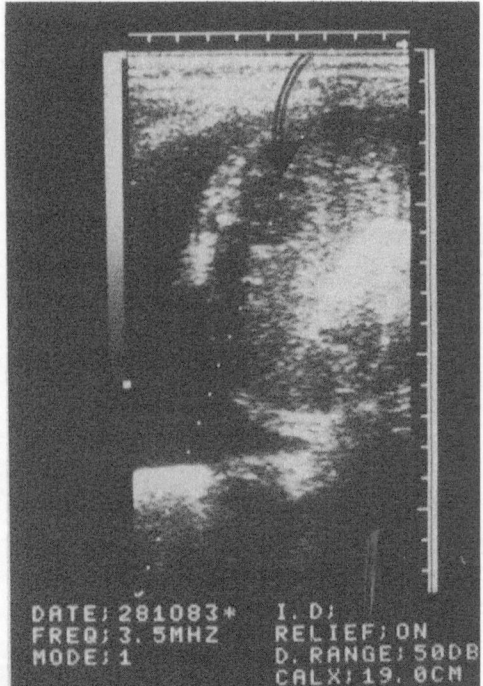

Fig. 2-2. Prediction of needle path
The electronically generated dotted marker line transects the small target. In this case a small abscess in the spleen (note: left pleural effusion).

17

enabling the needle to enter the rectangular image field vertically or slightly angled from its top. The needle steering canal can also be detachable and mounted at the end of the transducer.

One of the advantages of real-time puncture monitoring is that compression of tissue and friction between the needle and the tissues along the tract is appreciated, so that any induced alteration in the distance to the target can immediately be taken into account (Fig. 2-3).

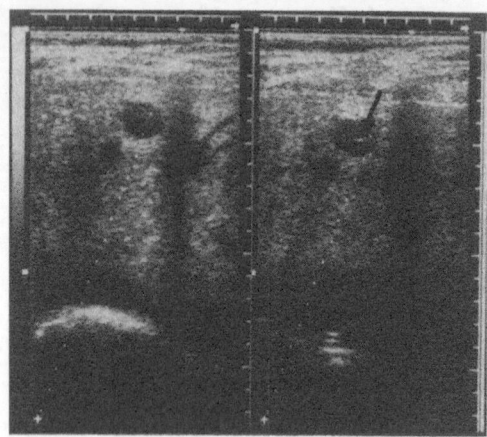

Fig. 2-4. Visualization of needle tip
In the right part of this double image, the echo from the tip of a fine needle (0.6 mm) is clearly seen inside a small solid hepatic mass lesion.

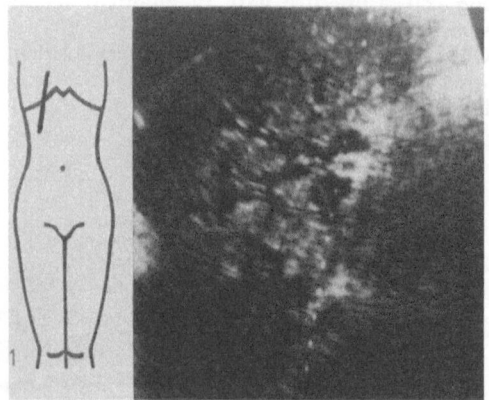

Fig. 2-3. Precise needle monitoring
In this case, where ultrasonically guided fine needle PTC is intended, real-time monitoring is mandatory, as the target is tiny and relatively deeply situated. The line of sight transects (from above): a bile duct, a branch of the hepatic artery and a branch of the right portal vein. While aspirating bile and injecting X-ray contrast, the needle tip must be retained inside the lumen of the bile duct.

If bending of the needle should occur, this can either be seen directly (curved needle track on screen) or, more often indirectly by lack of needle tip echo inside the lesion due to deviation of the needle out of the image plane (Fig. 2-4). Physiological movements of targets (e.g. pulsatile, respiratory, fetal movements) are evaluated before needle introduction, to determine the optimal instance of rapid needle introduction, if the movements cannot be controlled by other means (e.g. short voluntary patient apnoea).

Commonly used needles

Various purposes require different needles – different regarding shape and diameter – while the length is determined by the canal in the needle steering device and the desired depth of puncture. Needles with a length of 20–25 cm will be sufficient. Fine needles having outer diameters of 0.6 mm should be used when puncturing solid lesions, where the possibility of malignancy exists and also whenever they can be expected to fulfill the purpose, because of their minor tissue damage and thereby minor inherent risk of causing complications such as bleeding.

The fine needles may be very flexible and in order to stabilize their route, an outer guide needle with an o.d. of 1.2 mm, 10 cm long, is first introduced through the superficial layers, e.g. the abdominal wall. This needle will also act as a sheath and probably reduce the risk of seeding tumor cells in the needle tract when the fine needle is passed in and out the suspect lesion more than once to obtain sufficient material.

The necessary vacuum for fine needle aspiration biopsies is most easily obtained by means of a one hand aspiration handle mounted with a disposable 10 ml syringe.

In the last few years, fine needles with cutting capability have become available. From the tiny, slim tissue cores histological diagnoses can be obtained without the necessity of using large bore cutting needles (Figs. 2–5, 2-6).

For puncturing and emptying of fluid-containing, most certainly benign

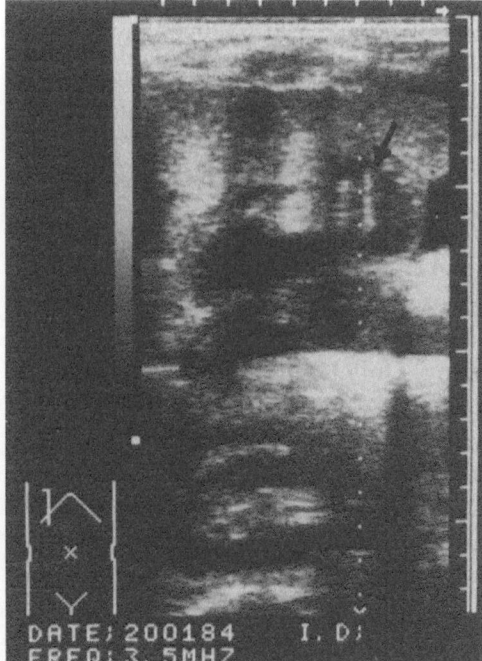

Fig. 2-6. **Fine needle cutting biopsy situation** The cutting needle, Surecut® 0.8 mm, TSK, Japan, is visualized inside a tumor at the border of the right liver lobe. Ascites is also present.

Fig. 2-5. **Modified fine-caliper Menghini needle** The distal tip of the needle with trocar and after withdrawal of trocar. This type of needle provides suction and cutting of the biopsy specimen (Surecut®, TSK, Japan).

lesions, needles with larger lumens are practical. A common lumbar cannula with an o.d. of 1.2 mm is routinely used (Fig. 2-7). This will also allow passage of a guide wire, should the need for catheter drainage a.m. Seldinger arise.

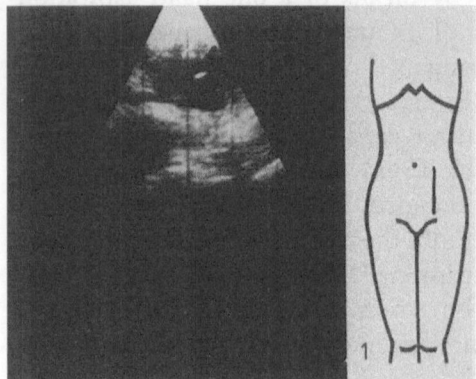

Fig. 2-7. Aspiration of fluid
In this case of an abscess in the left iliac fossa, a lumbar cannula with an o.d. of 1.2 mm was used for diagnostic aspiration and possible emptying. In such cases the strongly reflecting needle tip is easily kept in position during aspiration and cavity collapse.

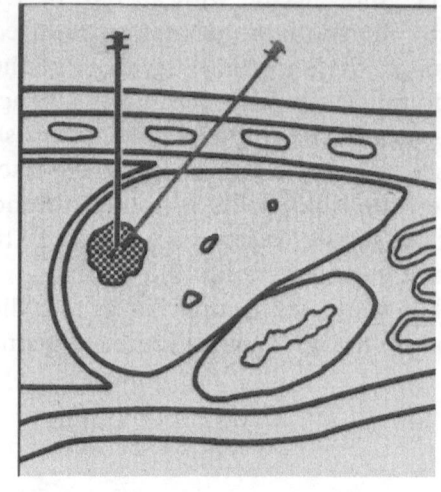

Fig. 2-8. Avoid transpleural and transpulmonal puncture

Safety precautions

When the ultrasonic examination has disclosed a puncture target, the puncture should be performed in the same setting. Only large diameter needle liver and kidney biopsies require prior fasting, compatible blood in the bank and post-puncture bedrefinement and observation. In all other cases the puncture may be undertaken on an out-patient basis and all pre- and postpuncture care required is to inform the patient about what will be undertaken.

If any clinical suspicion of increased bleeding tendency exists, the coagulation parameters should be checked. Moreover puncture should be avoided if suspicion of aneurysm or pheochromocytoma should be relevant. Transpleural and especially transpulmonal puncture is also avoided in puncture of abdominal targets (Fig. 2-8). Whenever puncture is required, the optimum needle route – often the shortest – is chosen. One should not hesitate to tra-verse stomach, intestine or urinary bladder to reach a target, as comprehensive experience has shown this to be safe (Fig. 2-9).

During the procedure, the patient should be continuously reassured and informed of what is happening, and

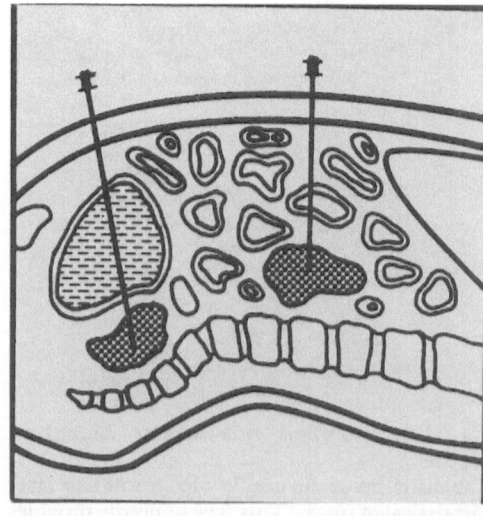

Fig. 2-9. Urinary bladder and intestinal loops may be traversed during aspiration

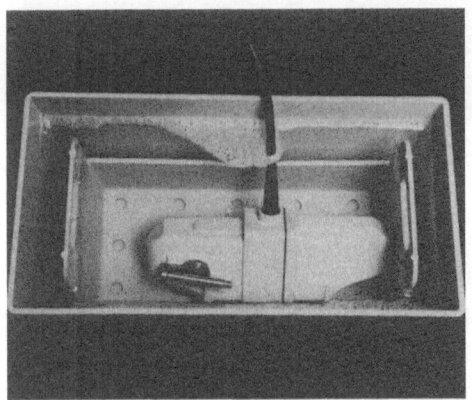

Fig. 2-10. Fluid sterilization
The transducer is sterilized by immersion in a suitable solution, e.g. chlorhexedine in alcohol. It may be necessary to consult the manufacturer, to secure which chemicals the transducer will withstand and which portion of it is fluid-proof.

should avoid deep breathing to increase the chance of precise puncture at the first attempt and to avoid the risk of organ damage.

The skin can be draped or washed and local anesthetic infiltrated. A sterile contact medium is applied and the transducer either wrapped in sterile foil or fluid sterilized, since it will not withstand autoclaving and since gas sterilization may be too time consuming (Fig. 2-10).

Performance of fine needle aspiration biopsy

In fine needle puncture, first the outer supporting needle is introduced through skin and superficial layers. This will allow for 2–4 passes of the fine needle, which is further advanced into the target.

In puncture under dynamic monitoring no mounted needle stop is necessary. Prior to the introduction of the fine as-

piration needle, its lumen has been checked for patency and connected to the vacuum-producing 10 ml syringe in the aspiration handle (Fig. 2-11).

Both gloved hands of the examiner are in use; one is supporting the transducer and the other steering the handle and thereby the needle.

When the needle tip has been placed,

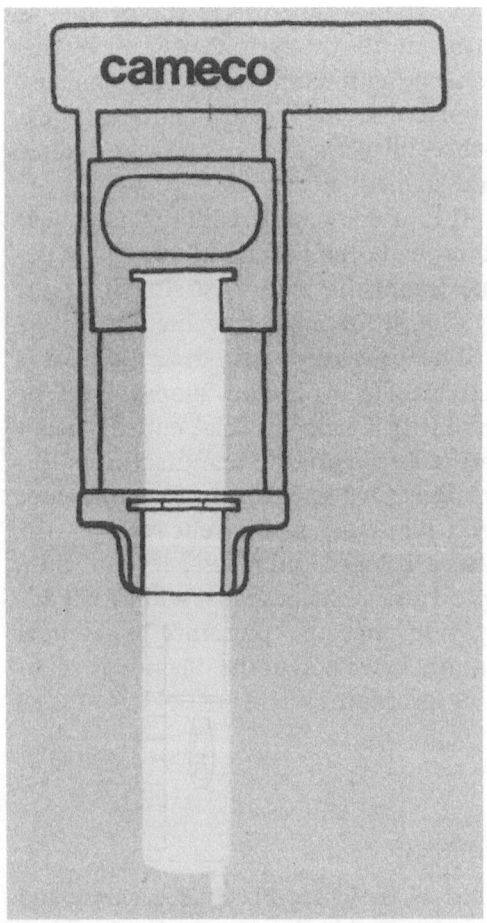

Fig. 2-11. A one-hand aspiration handle
This aspiration piston, manufactured from steel and aluminium, is sterilized together with the puncture-transducer, and will accept a disposable syringe with a capacity of 10 ml. Full retraction will provide an adequate vacuum for a long 0.6 mm aspiration biopsy needle (Cameco AB, Täby, Sweden).

21

full suction is applied and the needle moved 1 cm or so back and forth 3 or 4 times to loosen cell clusters, which will be retained in the lumen of the needle, while the content of fluid lesions or very bloody lesions tends to reach the syringe. Then the vacuum is neutralized by slowly releasing the handle, while the needle is still inside the lesion, to avoid superfluous mixing with aspirated material from outside the lesion. To remove the specimen from the needle, an assistant fills the syringe with air by disconnecting it from the needle. After reconnection, the needle content is expelled on glass slides, smeared, fixated and stained (Fig. 2-12).

It is often possible with the naked eye to evaluate the slides for content of tissue fragments before passing it to the cytologist for further evaluation.

The procedure just described for a fine needle aspiration biopsy may be modified if using special cutting needles, according to the manufacturers' directions for use. After some experience with puncture in different regions, the ultrasonologist will reach a level of skill, speed and gentleness that will justify his opinion that the puncture is a most natural extension to the diagnostic scanning procedure.

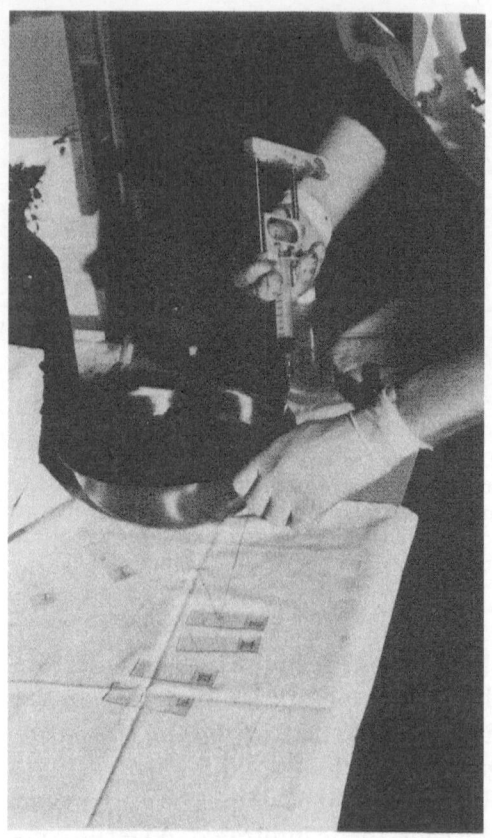

Fig. 2-12. Expulsion of aspirated material
The droplets containing the cell-clusters are expelled on glass slides and smeared. The following handling procedures are variable, according to staining method, etc.

References

Holm H H, Jensen Fl. Guided biopsy and drainage with ultrasonography. In: Margulis A R, Burhenne H J, eds. Alimentary tract radiology, vol. 2, 3rd ed. St. Louis: Mosby, 1983: 2374.

Holm H H, Kristensen J K (eds). Ultrasonically guided puncture technique. Copenhagen: Munksgaard and Baltimore: University Park Press, 1980.

Pedersen J F. Percutaneous puncture guided by ultrasonic multitransducer scanning. *J Clin Ultrasound* 1977; 5: 175.

Saitoh M, Watanabe H, Ohe H, Tanaka S, Itakura Y, Date S. Ultrasonic real-time guidance for percutaneous puncture. *J Clin Ultrasound* 1979; 7: 269.

CHAPTER 3

Handling of aspirated material

Sven Grønvall

A correct diagnosis can often be made based on the clinical setting and the sonographic study alone. In many cases, however, ultrasonically guided needle aspiration is essential to obtain a final diagnosis as precisely, rapidly and gently as possible.

A correct handling of the aspirated material involves many disciplines (Fig. 3-1) and requires a skilful and imaginative sonographer. The consistency of the puncture target, the character of the aspirate and the mascroscopic appearance often add significant information about the nature of the pathological mass, especially in cystic lesions, e.g. pus in abscesses, blood in hematomas etc.

Evaluation of aspirate

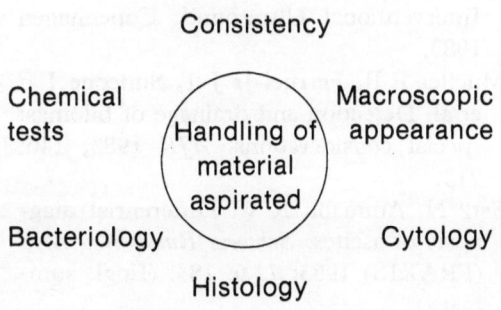

Fig. 3-1. Evaluation of aspirate

The evaluation of the aspirated material may form the basis for immediate therapy as in abdominal fluid collections and abscesses (see chapter 25).

Solid lesion

When the puncture target is solid a fine needle aspiration biopsy is performed. If necessary, an experienced cytologist should be able to give a preliminary result within 30 min. In most cases cytologic examination can differentiate between a malignant and a benign lesion. For further specification of the cells, cytochemical and immunological procedures may be applied. Histological micro-core biopsies can be obtained with special fine, cutting needles. This technique, which has advantages as well as disadvantages, is described in chapter 6.

If a focus of infection is suspected, the aspirated material should always be examined bacteriologically. Aerobic as well as anaerobic cultures are necessary.

Cystic lesion

Depending on the clinical history and findings, the macroscopic appearance of the aspirate and the localization of a fluid-filled lesion, more specific biochemical tests may be applied (Table 3-1). Many of these tests, which can be performed qualitatively or semiquantitatively using Multistix SG® (Ames Compagny Division of Miles Lab. Ltd. Stoke Poges, Slough SL2 4LY, England), are often helpful in the differentiation of various fluid collections. Thus a high creatinine content is found in urinomas, high amylase content in pancreatic pseudocysts and high bilirubin content in cholascos. In patients with ascites or pleural effusion, the specific gravity of the aspirate may allow differentiation between a transudate and exudate since the specific gravity is higher in an exudate (1.016–1.020) than in a transudate (1.010–1.012). The transudate may indicate cardio-renal failure and the exudate possible inflammatory and/or carcinomatous involvement.

From a simple renal cyst clear yellowish fluid containing protein and glucose is aspirated. Additional laboratory tests show low or normal lactid dehydrogenase, no fat and negative cytology, whereas increased contents of lactid dehydrogenase and cholesterol in a renal cystic lesion suggest malignancy, when inflammation and intracystic hemorrhage can be excluded.

T. Livraghi and coworkers from Italy showed raised values of potassium and chlorine in fluid from hydatid cysts. In areas where ecchinococcosis is endemic these electrolyte data could be of diagnostic value when a hydatid cyst is punctured percutaneously.

Examination of material obtained by ultrasonically guided puncture often provides for a rational approach, speeding up diagnostic evaluation of the patient in daily clinical work. A number of the diagnostic tests can be performed by the sonographer in conjunction with the ultrasound examination.

Table 3-1. Biochemical tests

Test	Elevated in
Creatinine	Urinoma
Amylase	Pancreatic cyst
Protein (+)	Exudate (Pleural, ascites)
	Lymphocele
Bilirubin (+)	Biloma
Protein, Glucose (+)	Simple renal cyst
LDH, Cholesterol	Malignant renal cyst
Potassium, Chlorine	Hydatid cyst
Specific gravity (+)	Exudate (Pleural, ascites)
pH (+)	

(+) indicates when Multistix SG® can be used.

References

Ames reagent tests. Technical information manual, 1982.

Kleist H, Jonsson O, Lundstam S, Nauclér J, Nilson A E, Pettersson S. Quantitative lipid analysis in the differential diagnosis of cystic renal lesions. *Br J Urol* 1982; 54: 441.

Livraghi T, Bosoni A, Giordano F, Lai N. Diagnosis of hydatid cyst by percutaneous aspiration: value of electrolyte determinations. Third International Congress on Interventional Ultrasound, Copenhagen 1983.

Mueller P R, Ferruci Jr J T, Simeone J F et al. Detection and drainage of bilomas: special considerations. *AJR* 1983; 140: 715.

Satz N, Ammann R W. Differential diagnosis of ascites. *Schweiz Rundschau Med* (PRAXIS) 1983; 72,6: 183. (Engl. summary).

CHAPTER 4

General principles of fine needle aspiration cytology

Grete Krag Jacobsen

Cytology is concerned with the evaluation of cellular properties based on interpretation of the cellular appearance. Prerequisites for correct interpretation are the procurement of sufficient and representative material prepared with a minimum of artefacts.

Preparation of the smear

In order to obtain optimal material from a solid lesion the aspiration should be performed from different areas within the lesion, as many malignant tumors contain foci with necrosis, hemorrhage, cysts and fibrosis in addition to viable tumor tissue. The aspirated material is expelled onto glass slides in single drops which are spread immediately as shown in Fig. 4-1. Using this method, several specimens will be available from each lesion. Material from cysts is centrifuged before being placed on glass slides and spread as described above.

Fixation and staining of the smear

The specimens may be wet fixed immediately or air dried. Wet fixation, which is done while the surface of the slide is still wet, is usually performed in

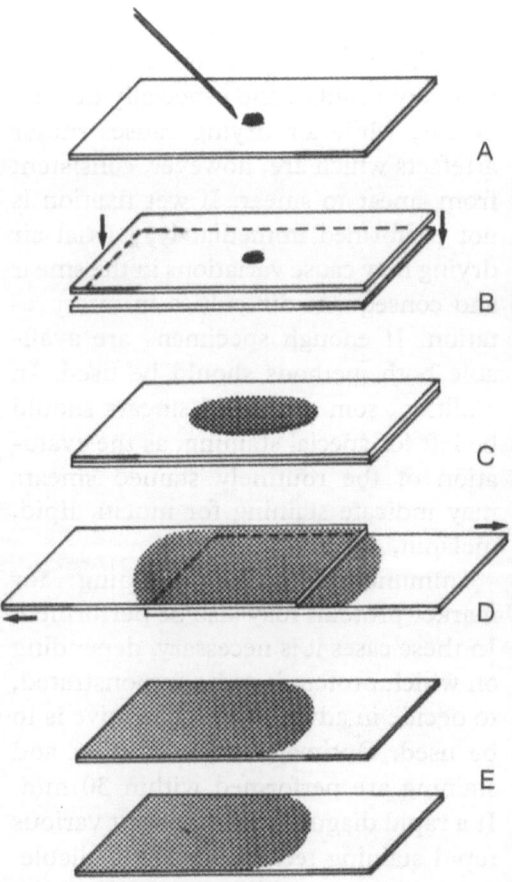

Fig. 4-1. Spreading of the aspirated material
A small drop is expelled from the needle onto a clean glass slide (a). Another glass slide is placed on the first one (b) whereby the drop spreads between the slides (c), which are pulled gently in opposite directions (d), resulting in 2 specimens with the aspirated material spread in a thin film on each slide(e).

95% ethanol. Some prefer a spray fixative as used for vaginal smears. Wet fixation is usually followed by the Papanicolaou staining. The air dried specimens, which may be stained at once or later (days or even weeks), are fixed in methanol during the first step of the staining procedure. The staining is usually performed according to the May-Grünwald-Giemsa technique which is the routine staining used by hematologists for smears of blood and bone marrow. Wet fixation gives optimal preservation of cellular, and especially nuclear details, while air drying causes minor artefacts which are, however, consistent from smear to smear. If wet fixation is not performed immediately, partial air drying may cause variations in the smear and consequent difficulties in interpretation. If enough specimens are available both methods should be used. In addition, some air dried smears should be left for special staining, as the evaluation of the routinely stained smears may indicate staining for mucin, lipid, melanin, etc.

Immunocytochemical staining for marker proteins may also be performed. In these cases it is necessary, depending on which protein is to be demonstrated, to decide in advance which fixative is to be used. Optimal routine fixation and staining are performed within 30 min. If a rapid diagnosis is important various rapid staining techniques are available. A rapid hematoxylin and eosin technique or a modified Papanicolaou staining may be carried out in less than 2 min. These stainings may also be used for evaluation of the quantitative and qualitative adequacy of the smear. If necessary the smear may be restained with the ordinary technique.

Cytologic evaluation of the smear

Since the indication for fine needle aspiration biopsy in most cases is the presence of a solid mass lesion, the most frequent question is whether or not the lesion is malignant. In cytology the most important signs of malignancy such as invasiveness and metastases can only rarely be demonstrated. The recognition of malignant cells in smears is based on the cytologic criteria of malignancy. The malignant cells show changes of nuclei and cytoplasm when compared with normal cells. Thus the usual cytologic criteria of malignancy include:

1. Variation in size and shape of cells;
2. Variation in size and shape of nuclei;
3. Changes in amount and structure of nuclear chromatin;
4. Increase in size and number of nucleoli;
5. Increase in number of mitoses and presence of abnormal mitoses;
6. Changes of nuclear polarity;
7. Increase of nucleo-cytoplasmic ratio;
8. Change in cytoplasmic structure and staining properties.

In addition, aspirates from malignant lesions usually show a high cellularity compared to aspirates from benign lesions or normal tissue. Tumor cells are also often present as clumps and thick clusters as well as single cells on a cellular background with necrotic material, while normal cells, due to normal cellular cohesion, occur in single layer sheets with a regular pattern, together with only a few single cells on a "clean" cellular background.

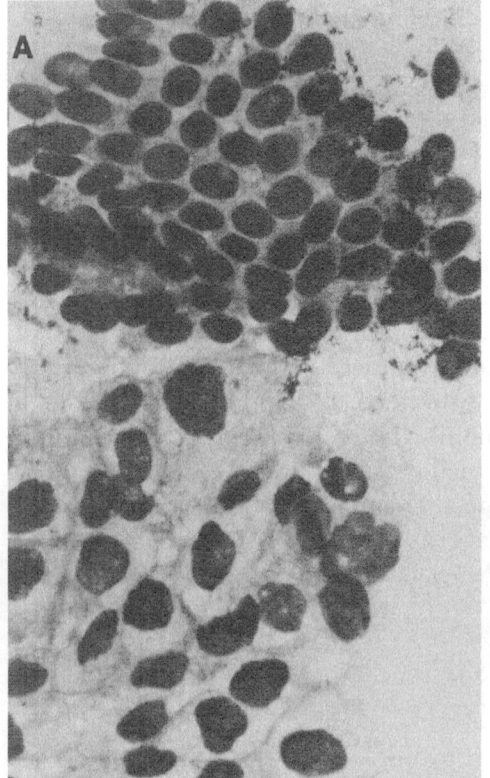
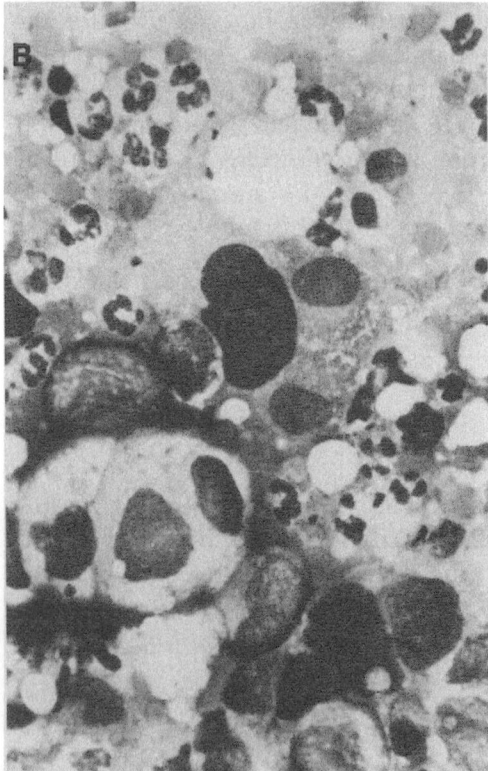

Fig. 4-2. Fine needle aspiration from the pancreas

A. Normal ductular cells (top) arranged in a monolayer sheet with a regular pattern and with good cohesion between the cells next to a sheet of moderately differentiated tumor cells (below) with moderate variation of size and shape of cells and nuclei as well as with variation of nuclear polarity and cellular arrangement.

B. Clusters and single cells from a poorly differentiated tumor exhibiting many signs of malignancy (variation of size and shape of cells and nuclei, coarse structure of nuclear chromatin, large nucleoli, increased nucleo-cytoplasmic ratio, etc.) present in the smear in addition to necrotic material and inflammatory cells.

The cytologic criteria of malignancy are all useful, but none or very few are absolute. However, as a rule only some of these criteria are encountered in the smear from a malignant lesion. The presence of at least two of the criteria is necessary to identify a cell as malignant. The interpretation of malignancy will be more reliable if the changes are found in many cells, although a few cells with strong characteristics may suffice for a correct diagnosis.

It is important to be aware of the great variation in appearance of malignant cells from organ to organ as well as from tumor type to tumor type. Indeed, the same diagnostic criteria, when applied to different organs, may result in fatal diagnostic failure. This is one of the reasons why the smear should at least be accompanied by information on the precise site of the puncture and the sex and age of the patient. Many difficulties in cytologic interpretation can be overcome with experience, and many diagnostic pitfalls can be avoided when the cytologist has a broad knowledge of histopathology.

The difficulties in cytologic interpretation which clinicians also have to be aware of may be summarized in two main categories, as follows:

1. Some malignant cells do not exhibit the usual cytologic changes of malignancy; and
2. Some normal cells or cells from benign lesions imitate changes of malignancy.

Examples of the former are highly differentiated tumors, such as carcinomas of the kidney, the prostate and the salivary glands, and endocrine tumors, such as carcinomas of the thyroid and carcinoid tumors. The most important examples of the latter are cellular changes caused by inflammation and regeneration.

Thus, the clinician should always consider the cytologic diagnosis in the context of the total amount of data available about the patient, taking into account sampling error as well as error of interpretation. Here the opinion of F. Stuart, one of the pioneers of cytology should be considered – an opinion which is as valid today as it was 50 years ago: "Diagnosis by aspiration is as reliable as the combined intelligence of the clinician and the pathologist makes it."

References

Esposti P L, Franzén S, Zajicek J. The aspiration biopsy smear. In: L. G. Koss, ed. Diagnostic cytology and its histopathologic basis. 2nd edn., Philadelphia: J. B. Lippincott, 1968.

Frable W J. Fine-needle aspiration biopsy: A review. *Hum Pathol* 1983; 14: 9.

Pak H Y, Yokota S B, Teplitz R L. Rapid staining techniques employed in fine needle aspirations. *Acta Cytol* 1983; 27: 81.

Sachdeva R, Kline T S. Aspiration biopsy cytology and special stains. *Acta Cytol* 1981; 25: 678.

Zajicek J. Aspiration biopsy cytology, part I. Cytology of supradiaphragmatic organs. Vol. 4, Monographs in Clinical Cytology. Basel: Karger, 1974.

Zajicek J. Aspiration biopsy cytology, part II. Cytology of infradiaphragmatic organs. Vol. 7. Monographs in Clincial Cytology. Basel: Karger, 1979.

CHAPTER 5

Electron microscopy of ultrasound guided fine needle biopsies

Maxwell Sehested, Niels Juul & Bo Hainau

Diagnostic electron microscopy (EM) is used in surgical pathology as a supplementary procedure to light microscopy (LM) in the classification of certain tumors. In this context EM can be of decisive importance to the correct clinical approach, be it the choice of operative procedure or the application of the appropriate chemotherapeutic regimen.

The following list presents groups of tumors in which the LM histological appearance can be indistinguishable but where EM is often able to supply a definitive diagnosis by demonstrating ultrastructural markers of tumor cell differentiation.

1. Small round cell tumors

This group of tumors includes small cell malignant lymphomas, small cell carcinoma of the lung (oat cell carcinoma), small cell neuroendocrine tumors (APUDomas), Ewing's sarcoma and rhabdomyoblastoma.

2. Large cell undifferentiated tumors

In this group of tumors EM plays a role in distinguishing between anaplastic carcinoma, sarcoma, large cell malignant lymphoma and malignant amelanotic melanoma. In light microscopical undifferentiated carcinomas, EM is often able to determine whether the tumor is an adenocarcinoma or a squamous cell carcinoma.

3. Spindle cell tumors

EM is used to differentiate between spindle cell squamous cell carcinoma and spindle cell sarcomas such as neurogenic tumors, leiomyosarcomas and fibrosarcomas.

It is, however, necessary to stress the limitations of EM in surgical pathology. These can be summarized as follows:

1. EM is not a good method of determining whether a given tumor is benign or malignant.
2. In the great majority of epithelial neoplasms (adenocarcinomas and squamous cell carcinomas) EM supplies no additional diagnostic information to that acquired by LM.

Method

A simple method is aspiration through the usual 0.6 mm needle using the same introduction and suction techniques as for LM (as described in Chapter 2). EM has been applied by several authors to fine needle aspirated material from tumors in various sites with good results. We have used the following procedure in aspiration biopsies:

1. A 0.6 mm needle is used.
2. Aspiration for LM is performed as usual (as described in Chapters 2 and 4).
3. The final aspirate is ejected directly into a small plastic conical tube (e.g. Micro Test Tubes, polypropylene, 1.5 ml, Milian Instruments S.A., Geneva) containing a suitable EM-fixative (e.g. 70% Karnovsky's fixative) and *not* onto a glass specimen slide.
4. After fixation for 2–4 h the suspension is centrifuged for 30 min at 3000 rpm, whereby a pellet is formed.
5. The pellet is cut out of the plastic conical tube and further processed as if it were a surgical tissue biopsy.

In fine needle histological biopsies with the Surecut® needle (as described in Chapter 6) the procedure is the same in that the obtained tissue core is immediately ejected into the plastic conical tube containing fixative. With this procedure, however, centrifugation is unnecessary as the tissue cores are visible to the naked eye and can be processed directly.

After the material for EM has been embedded in Epon, 1 μm sections are cut and stained with toluidine blue. This enables identification of tumor cells by LM which must correspond in appearance to those examined by LM either cytologically, or histologically if a

Surecut® needle has been used. Having thus confirmed by LM that the tumor tissue obtained for EM is representative, ultrathin sections are cut and examined by EM.

Results

Over a period of 1 year we have applied EM on a total of 38 ultrasound guided fine needle biopsies (FNB) in patients using the following indications.

1. Suspected metastases from unknown primary sites.
2. Suspicion of primary retroperitoneal tumors.
3. Suspicion of sarcoma, malignant lymphoma or neuroendocrine tumors.

33 FNB were aspirates and 5 FNB were Surecut® biopsies. In 6 FNB (16%) material for EM was either too necrotic for evaluation or devoid of tumor cells. In 10 cases (26%) the EM diagnosis was identical to that of LM, while in 22 cases (58%) EM was able to provide additional morphological information and thus supply a more accurate tumor diagnosis. We also analyzed these 22 cases with respect to whether the additional EM diagnosis had contributed significantly to the clinical management of the patients. This was so in 10 of the 22 cases or 26% of the total material. These 10 cases included 3 patients with a small cell carcinoma of the lung (oat cell carcinoma) metastatic to the liver, 1 patient with a retroperitoneal paraganglioma and 1 patient where a tentative LM diagnosis of malignant lymphoma was excluded in favor of carcinoma. The remaining 5 patients had metastatic tu-

mors which by LM were suspected of being endocrine (APUDomas). EM was able to verify this diagnosis in 3 cases and refute it in 2 cases.

We are therefore able to conclude that performing EM on ultrasound guided FNB was possible in a majority of cases (32/38) and of major clinical value in approximately a quarter of cases (10/38). Based on our experience with EM on ultrasound guided FNB and on EM in surgical pathology in general we recommend the following indications:

Indications

1. Patients with metastatic liver or retroperitoneal tumors when a primary lung tumor is suspected clinically or radiographically. This enables the therapeutically important diagnosis of small cell lung cancer to be established or disproved.

2. Patients with primary retroperitoneal tumors (other than pancreatic), tumors in unusual sites or with atypical clinical histories. This group of patients will thus include those suspected clinically or ultrasonically of having sarcoma, metastatic malignant melanoma, malignant lymphoma, or neuroendocrine tumors.

3. Patients where the LM diagnosis on ultrasound guided FNB is unable to classify the tumor and thus raises the suspicion of sarcoma, malignant melanoma, malignant lymphoma, or neuroendocrine tumor. In such cases a repeat ultrasound examination and FNB for EM evaluation is warranted for the appropriate clinical management.

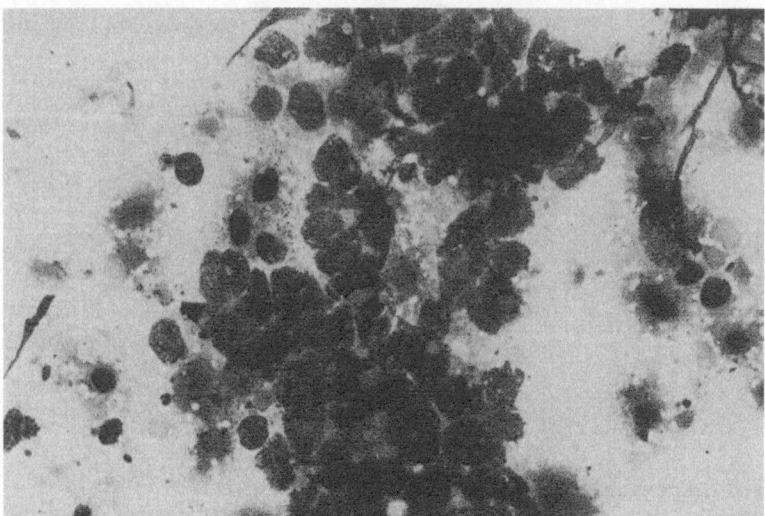

Fig. 5-1. Light microscopy, suspected liver metastasis
Light microscopy of cytological specimen from a suspected metastasis to the liver. The primary tumor was unknown. The tumor cells are undifferentiated with fragile nuclei, indistinguishable nucleoli and scanty cytoplasm. The light microscopical appearance suggests metastasis from a small cell carcinoma of the lung but is unable to give a definitive diagnosis, as other small round cell tumors must be considered. (May-Grünwald-Giemsa, × 250).

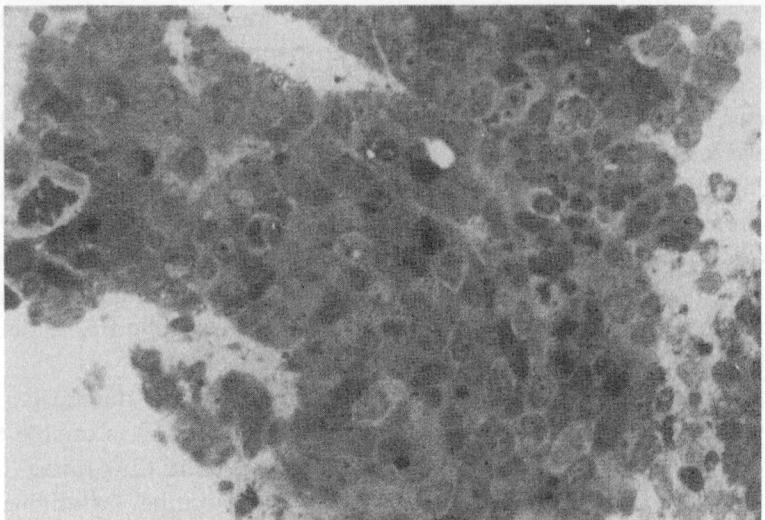

Fig. 5-2. Light microscopy, suspected liver metastasis
Light microscopy of 1 μm Epon section from an aspirate of the same lesion as
above but processed for EM. Clumps of undifferentiated small round tumor
cells corresponding to the cytological appearance. (Toluidine blue, × 250).

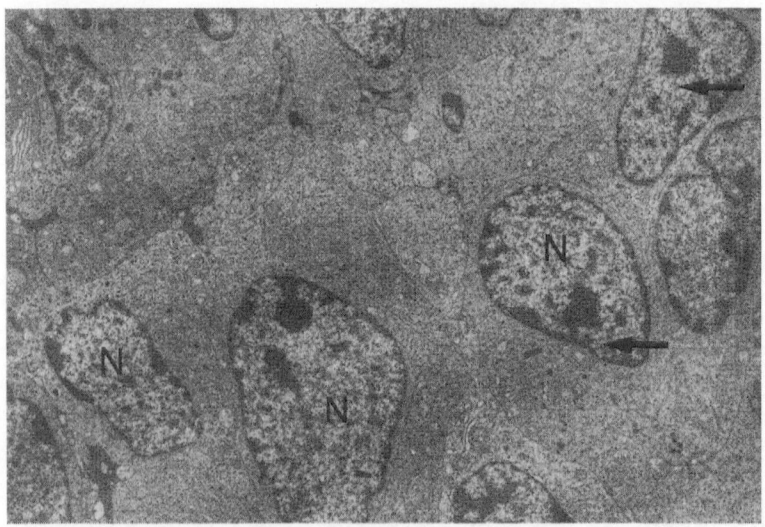

Fig. 5-3. Light microscopy, suspected liver metastasis
Low power electron photomicrograph demonstrating the same tumor cells with
small nuclei (N), inconspicuous nucleoli (arrows) and sparse cytoplasm. (Primary magnification × 2000).

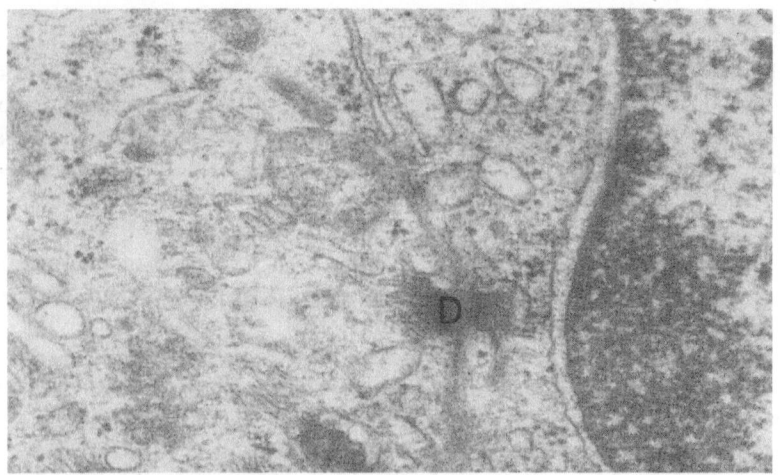

Fig. 5-4. Electron microscopy, epithelial tumor
High power electron photomicrograph from the same case demonstrating a desmosome (D) attaching two tumor cells. The presence of desmosomes signifies that the tumor is of epithelial origin and thus a tentative diagnosis of malignant lymphoma or sarcoma can be discarded. (Primary magnification × 12000).

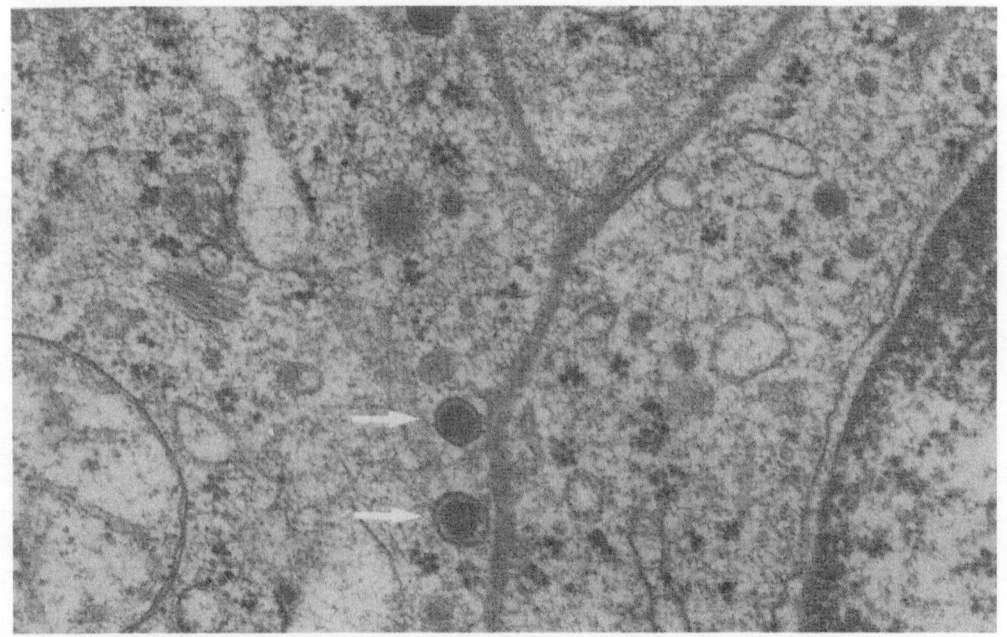

Fig. 5-5. Electron microscopy, endocrine tumor
High power electron photomicrograph showing a few dense core granules (arrows) with a diameter of 150 nm signifying an endocrine origin of the tumor. In this case it was thus possible to give a definitive diagnosis of metastatic small cell carcinoma of the lung using supplementary electron microscopy. (Primary magnification × 16000).

References

Ghadially F N. Diagnostic electron microscopy of tumours. London: Butterworths, 1980.

Henderson D W, Papadimitriou J M. Ultrastructural appearances of tumours. A diagnostic atlas. Edinburgh: Churchill Livingstone, 1982.

Lichtiger B, Mackay B, Tessner C F. Spindle cell variant of squamous carcinoma. A light and electron microscopic study of 13 cases. *Cancer* 1970; 26: 1311.

Akhtar M, Ashraf Ali M, Owen E W. Application of electron microscopy in the interpretation of fine-needle aspiration biopsies. *Cancer* 1981; 48: 2458.

Hagelqvist E. Light and electron microscopic studies on material obtained by fine needle biopsy. *Acta Oto-Laryngol* 1978; (Suppl.) 354.

Kindblom L-G. Light and electron microscopic examination of embedded fine-needle aspiration biopsy specimens in the preoperative diagnosis of soft tissue and bone tumors. *Cancer* 1983; 51: 2264.

Nordgren H, Åkerman M. Electron microscopy of fine needle aspiration biopsy from soft tisue tumors. *Acta Cytol* 1982; 26: 179.

Sehested M, Francis D, Hainau B. Electron microscopy of trans-thoracic fine-needle aspiration biopsies. *Acta Pathol Microbiol Scand* (Sect. A) 1983; 91: 457.

Karnovsky M J. A formaldehyde-glutaraldehyde fixative of high osmolality for use in electron microscopy. *J Cell Biol* 1965; 27: 137A.

CHAPTER 6

Fine needle histological sampling

Søren Torp-Pedersen, Mogens Vyberg, Niels Juul & Maxwell Sehested

The caliber of a biopsy needle should be large enough to yield a satisfactory amount of tissue for the pathologist and small enough to be virtually without risk to the patient. In daily practice fine needle aspiration biopsy has proven satisfactory in the diagnosis of malignancy. The needles used are 22 or 23 gauge and a high sensitivity and specificity can be obtained with very few complications. The technique demands the cooperation of an expert cytologist, however, and has not gained acceptance in all centers. Attempts to obtain material for histological examination with fine needles have therefore been made.

In 1981, Isler et al. described a method for obtaining histological material using a 22 gauge cutting needle. The needle with stylet was inserted and the stylet was removed when the needle was just outside the biopsy area. A syringe with saline was attached and with suction the needle was advanced into the biopsy region with a drilling motion. The contents of the syringe were then expelled into a test tube which was sent to the cytology laboratory. Tiny tissue cores were thereafter taken from the saline for histological processing. The technique is acceptable, but is not suited for ultrasonic guidance because it requires two hands.

In 1983 Livraghi et al. reported their experience with "inclusion cytology". Instead of smearing the material from fine needle aspiration biopsy, the content of the needle was imbedded for histological processing. This technique allowed histological evaluation of a cytological sample.

Both Isler and Livraghi found that additional valuable information was obtained with histological sampling.

Histological sampling with a 23 gauge modified menghini needle

In 1983 we developed a technique for histological sampling using the 23 gauge Surecut® needle. The Surecut® needle is a modified Menghini needle uniting needle, stylet, and syringe (Fig. 6-1). The stylet is attached to the plunger of the syringe and moves with it as the plunger is retracted. A locking device keeps the plunger retracted, thereby maintaining the negative pressure.

The biopsy procedure is only a slight modification of the fine needle aspiration biopsy. No further precautions are required when this needle is used as the caliber of the needle is the same. An 18 gauge (1.2 mm) outer guide needle is introduced into the abdominal wall

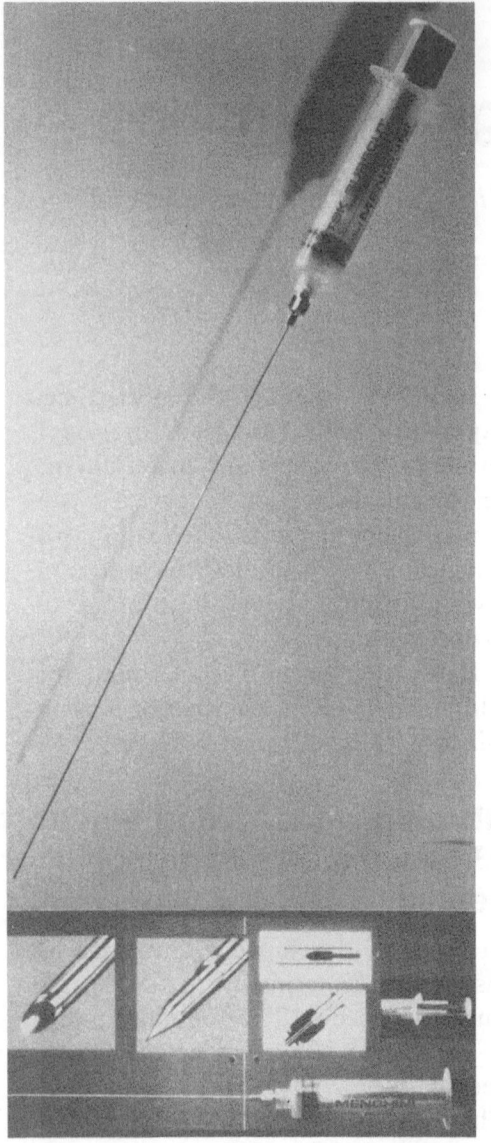

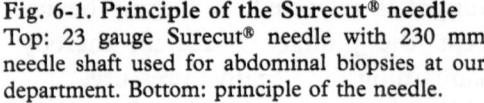

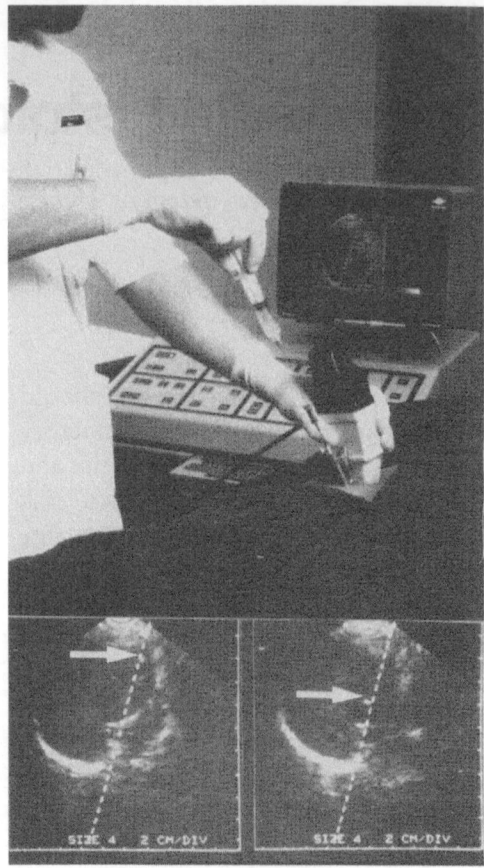

Fig. 6-2. Biopsy of a focal lesion in the liver
Top: when the tip of the needle is just outside the biopsy region the plunger of the syringe is retracted to prepare the needle for a cutting biopsy. Bottom: as the needle is advanced into the biopsy region the tip (arrows) is seen to move along the puncture line.

Fig. 6-1. Principle of the Surecut® needle
Top: 23 gauge Surecut® needle with 230 mm needle shaft used for abdominal biopsies at our department. Bottom: principle of the needle.

through the steering device mounted on the transducer. The Surecut® needle is then entered through the guide needle and the tip of the needle is seen to move along the puncture line on the screen. When the tip of the needle is just outside the target, the plunger of the syringe is retracted and the locking device keeps the plunger in this position (Fig. 6-2). The distal 3 cm of the needle is now a cutting needle with negative pressure. The needle is advanced into the target which is seen on the ultrasound image. The needle should now contain a tissue core approximately 3 cm long with a diameter of 0.4 mm. The needle is rotated to dislodge the tissue core and then removed.

The tip of the needle is brought into contact with a sterile piece of paper and the locking device is deactivated. The tissue core is expelled onto the paper as the needle is slowly pulled away from the initial point of contact (Fig. 6-3). Care should be taken that the tissue core is not coiled or fragmented.

Fig. 6-3. Tissue cores
Top: third tissue core being expelled onto sterile paper as the needle is slowly pulled away from the initial point of contact. Bottom: tissue cores are fixed in formalin still attached to the paper.

When the tissue has been expelled the needle is ready for another biopsy. Biopsies are taken from different places in the target and the tissue cores are placed on the paper as described above. The whole procedure is carried out with one hand.

Histological processing

When the 3 tissue cores have been expelled onto paper they are carefully straightened out with the needle. Still attached to the paper, they are fixed in neutral buffered formalin for about 4 h. After fixation the tissue cores are dislodged from the paper with a scalpel and placed between 2 pieces of gauze before dehydration and paraffin embedding. A drop of alcian blue on the tissue cores prior to dehydration aids the subsequent location of the cores in the paraffin block.

We routinely cut approximately 30 sections from each block which are placed on 6 slides. The following stains are used: hematoxylin and eosin, van Gieson (modified with alcian blue and sirius red), periodic acid-Schiff (PAS) and PAS after diastase digestion. Thus, 2 unstained slides are kept for special stains if needed, e.g. identification of argentaffin cells, amyloid, etc.

Experience with microcore biopsy

In 70 consecutive cases of suspected abdominal malignancy we have compared fine needle aspiration biopsy with fine needle histological biopsy. A 23 gauge fine needle without stylet was used for the aspiration biopsies as described in Chapter 2. The first 3 passes were made with the Surecut® needle followed by 3 aspiration biopsies. The cytological and histological material was evaluated by 2 different pathologists, who were given identical clinical information.

The distribution of the biopsies is shown in Table 6-1. In this table the number and distribution of the inadequate biopsies are also listed. In Table 6-2 and Table 6-3 the results of the biopsies are listed in terms of malignancy and non-malignancy. The results of the fine needle aspiration biopsies are typi-

cal. There are false negative results, but no false positives giving a predictive value of malignancy of 100%. The results of the Surecut® biopsies are very similar. There were more inadequate biopsies, but only one false negative giving very high predictive values of both positive and negative findings.

Retrieval rate in cytology and histology

Different tumor tissues vary in a number of ways which could influence the retrieval rate of fine needle biopsies. Fibrotic tumor tissue will yield a meager cellular result by aspiration while adequate tissue cores are easily obtained by the Surecut® needle (Fig. 6-4). On the other hand, highly cellular, soft, and partially necrotic tumors will provide abundant cytological material, but often only inadequate amount of tissue by the cutting biopsy. The latter probably ac-

Table 6-1. Distribution of biopsies

| Organ | | Insufficient material | |
		Histology	Cytology
Liver	32		
Retroperitoneum	13	2	1
Pancreas	11	1	1
Kidney	8	5	
Gastrointestinal	3		
Ovary	2		
Omentum	1	1	1
	70	9	3

Table 6-2. Fine needle aspiration biopsy of 70 abdominal lesions

			Malignant	Benign	Insufficient Material
Final	Malignant	52	47	3	2
Diagnosis	Benign	18		17	1
	Total	70	47	20	3

Predictive value of malignancy: $\frac{47}{47} = 100\%$ (92–100).　　Retrieval rate: $\frac{67}{70} = 96\%$.

Predictive value of non-malignancy: $\frac{17}{20} = 85\%$ (62–97).　　95% confidence limits in brackets.

Table 6-3. Fine needle histological biopsies from 70 abdominal lesions

			Malignant	Benign	Insufficient Material
Final	Malignant	52	45	1	6
Diagnosis	Benign	18		15	3
	Total	70	45	16	9

Predictive value of malignancy: $\frac{45}{45} = 100\%$ (92–100).　　Retrieval rate: $\frac{61}{70} = 87\%$.

Predictive value of non-malignancy: $\frac{15}{16} = 94\%$ (70–100).　　95% confidence limits in brackets.

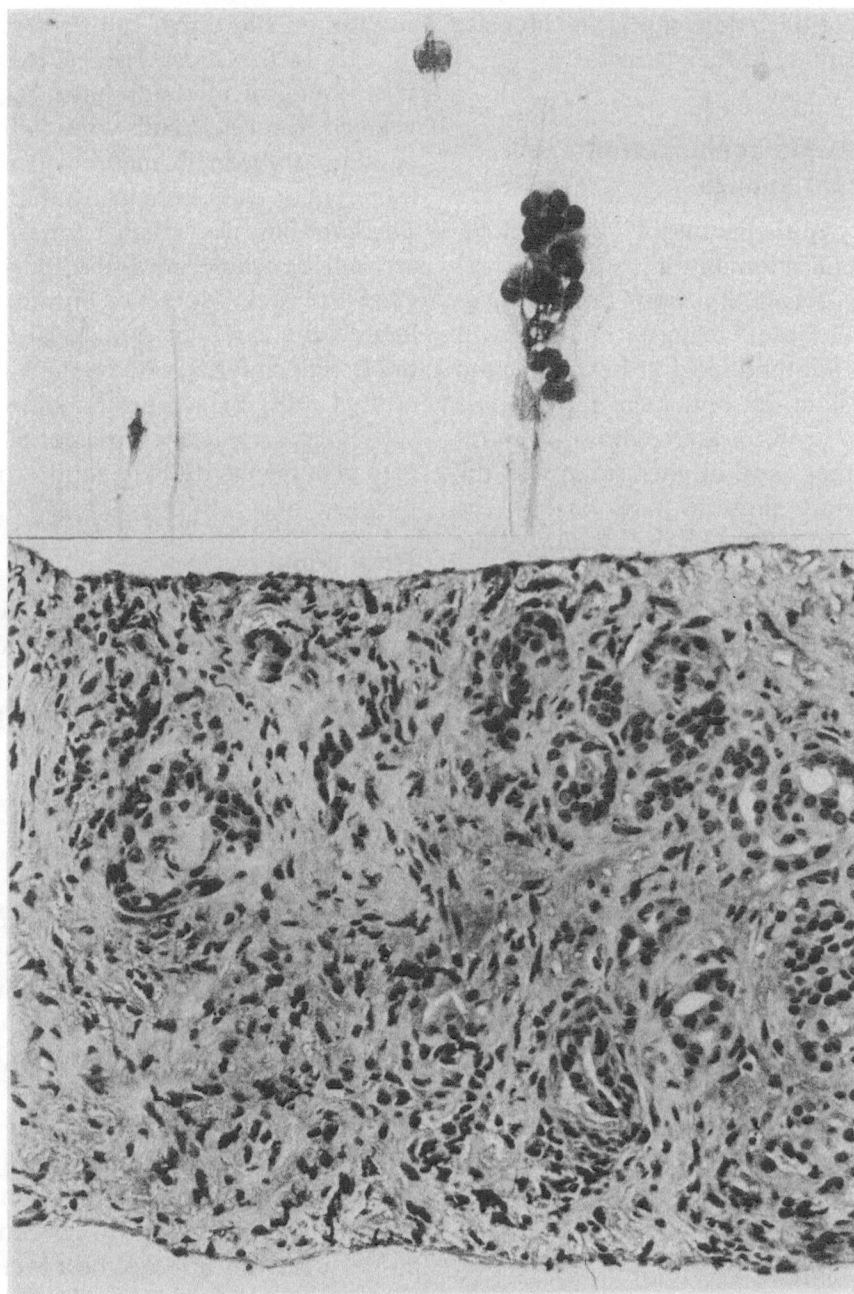

Fig. 6-4. Pancreatic biopsies
Top: sparsely cellular fine needle aspiration biopsy from the pancreas showing a small group of monomorphous epithelial cells. The cytological diagnosis is thus aspirate from the pancreas without malignancy (May-Grünwald-Giemsa, × 250). Bottom: Surecut® biopsy from the pancreas. Same case as top. The histological section demonstrates exocrine pancreatic tissue with severe fibrosis thus enabling a more precise diagnosis of the lesion. (Hematoxylin and eosin, × 250).

counts for the many inadequate biopsies from primary kidney tumors.

Microscopic evaluation of malignant tumors

In many epithelial tumors the histological classification of the Surecut® material will enable a more precise diagnosis. The great majority of malignant tumors encountered by the ultrasound specialist in the abdomen or retroperitoneum are adenocarcinomas. A number of these will be metastatic and the histological picture may enable the pathologist to suggest the location of the primary lesion (Fig. 6-5). This is only rarely possible in cytology. However, cytological examination can be more useful than histological examination in diagnosing malignant lymphomas and other mesenchymal tumors.

Microscopic evaluation of benign lesions

In benign pathological lesions, the histologic appearance of the tissue obtained by the Surecut® biopsy has distinct advantages over cytology. Histologic signs of fibrosis, edema and, to a certain extent, inflammation are undetectable at cytological examination. Histology may supply the explanation for the pathological process detected by ultrasound.

Special staining procedures

In the classification of tumors, as in the evaluation of other types of lesions, the use of special stains, including histochemical reactions, is important. It is a definite advantage to use these procedures on slides that can be compared directly to the slides stained routinely. Histochemical methods have been developed in surgical pathology for use on paraffin embedded material and can therefore be used without trouble on the Surecut® biopsies. Histochemical stains are, on the whole, unsuited for cytological aspirates. Moreover, in most centers all the cytological material is routinely stained for fear of overlooking abnormal cells. Blanching the slides with subsequent, application of special stains may give results difficult to interpret.

Time consumption

It is much less time consuming for the pathologist to examine serial sections of a histological biopsy than it is to look through 10–12 smears from an aspiration biopsy.

Advantages and limitations of Surecut® biopsies

The advantages of fine needle histologic sampling listed here will of course be more pronounced in centers without cytological experience. Jacobsen et al. have shown that ultrasonically guided fine needle aspiration biopsies of focal lesions of the liver have a higher diagnostic accuracy than ultrasonically guided large bore biopsies. They conclude that this is probably due to the multiple passes with the fine needle covering a larger volume of the liver, thereby providing a better chance of obtaining tumor cells. In our material the incidence of false negative diagnoses is not higher with histology, because multiple passes were also made with the Surecut® needle. The aspiration biopsy will probably

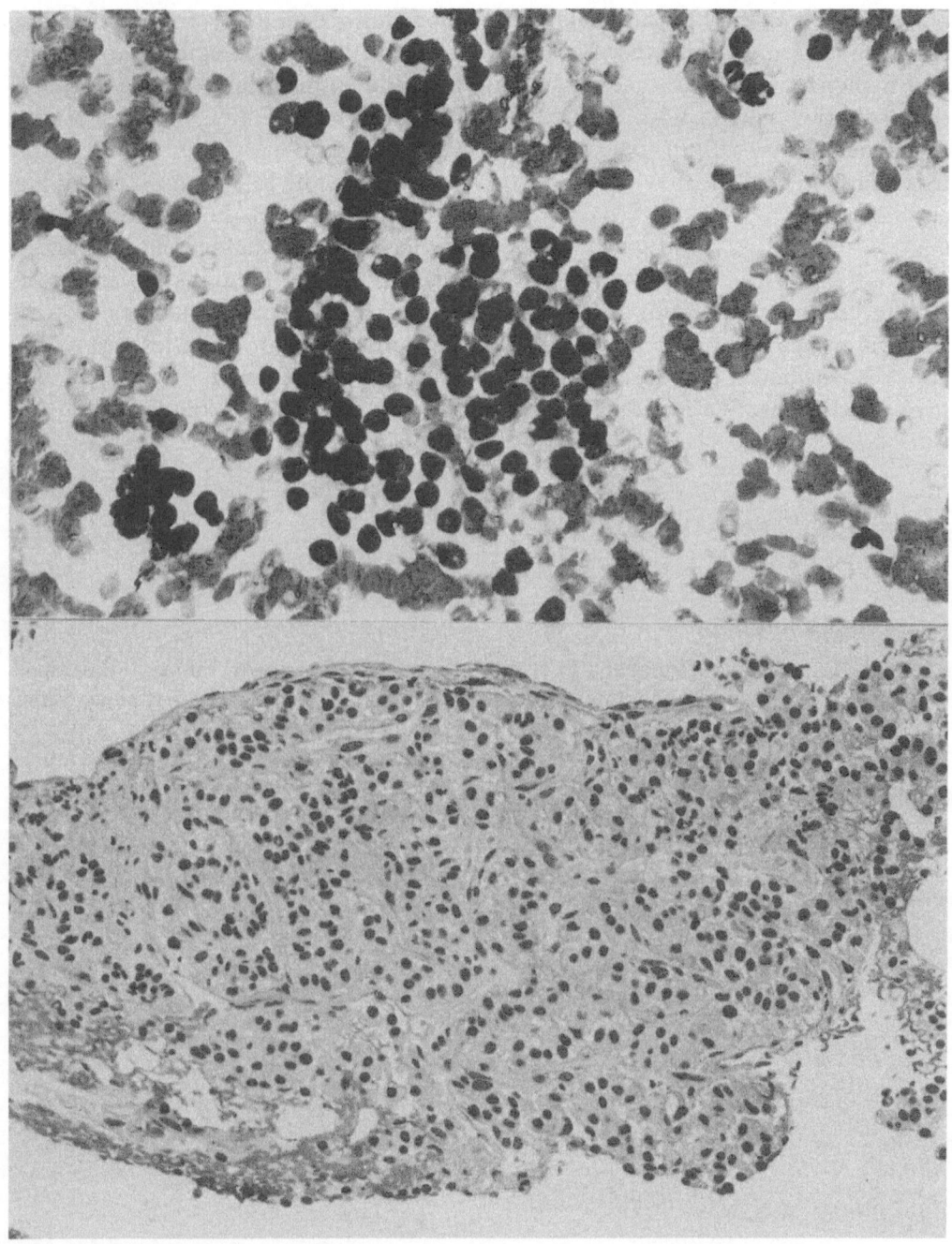

Fig. 6-5. Retroperitoneal tumor biopsies
Top: cellular fine needle aspiration biopsy from retroperitoneal tumor. The cells are small with regular round nuclei and suggest an endocrine neoplasm. (May-Grünwald-Giemsa, × 250). Bottom: Surecut® biopsy from the same case as top. The histological section shows the same cell type as in top, but also demonstrates the characteristic histological architecture of a paraganglioma with the cells arranged in nests or "Zellballen". (Hematoxylin and eosin, × 150).

always have a slightly higher diagnostic accuracy in terms of malignancy versus non-malignancy, because each pass with the aspiration needle with its up and down movements will cover a larger volume than the single stab with the Sure-cut® needle.

The technique of obtaining histological material with a fine needle as described above is new and it is too early to draw firm conclusions. We feel that the method shows definite promise in enabling a precise histopathological diagnosis to be made on material obtained by ultrasonically guided fine needle biopsy. However, further studies are necessary to determine its role *vis à vis* traditional aspiration cytology. Future comparative studies on the advantages and disadvantages of the two methods in various organs will allow conclusions as to whether histological biopsies can significantly supplement or even replace fine needle aspiration biopsy.

References

Isler RJ, Ferruci JT, Wittenberg J, et al. Tissue core biopsy of abdominal tumors with a 22 gauge cutting needle. *Am J Roentgenol* 1981; 136: 725.

Jacobsen GK, Gammelgaard J, Fuglø M. Coarse needle biopsy versus fine needle aspiration biopsy in the diagnosis of focal lesions of the liver. *Acta Cytol* 1983; 27: 152.

Livraghi T, Mitella M, Pilotti S, Ravetto C, Sangalli D, Solbiati L. Histology versus cytology in fine-needle biopsy. Third International Congress on Interventional Ultrasound. Copenhagen, 1983.

Torp-Pedersen S, Juul N, Vyberg M. Histological sampling with a 23 gauge modified Menghini needle. *Br J Radiol* 1984; 57: 151.

Puncture of focal liver lesions

Aksel Haubek

The ability of modern ultrasound equipment to display subtle parenchymal changes of the liver has contributed substantially to the role of sonography in the evaluation of liver disease. The primary application has been the detection of focal liver lesions, but a large number of publications and ongoing investigations indicate that sonography may gain increasing importance in the future classification of diffuse or parenchymal liver disease as well. Focal lesions of the liver are not synonymous with metastatic deposits and a wide variety of benign pathological conditions may present with a focal appearance on sonography. The frequency of benign focal liver lesions is not a new discovery for the pathologist, but it must be recognized by sonographers.

The sensitivity of sonography in the detection of focal liver lesions is poorly documented in the literature and carefully conducted prospective studies in this field are needed to demonstrate the accuracy, the precision and the utility of hepatic sonography. One explanation for the present situation may be found in the apparent methodological obstacles to prospective studies; another may be the convincing uncontrolled experience of the value and limitations of the method. In recent years there seems to have been a tendency towards replacing scintigraphy with sonography as the method for screening patients for secondaries in the liver. This change of attitude may be justified by an apparent higher specificity of sonography as well as by other virtues. On the other hand, this only more profoundly calls for conclusive prospective studies and hopefully the following depressing, but realistic statement concerning scintigraphy (Christensen & Rødby 1982) will not have to be applied to sonography at some future date: "Surprisingly – for a clinical tool used for nearly thirty years – it still seems too early to make firm conclusions regarding the diagnostic value of liver scintigraphy to disclose metastases in a surgical population."

The ultrasonically guided puncture is an important prerequisite for the specificity of hepatic sonography and much experience on the indications, the utility and the risks of this procedure has accumulated in recent years.

Technique

The fundamentals of the sonographically guided puncture and biopsy techniques are well established. A number of modifications may be utilized according to personal preference and new types of needle design involve current changes. Generally, dynamic equipment is recommended as is the use of needle attachment or puncture probes. Cystic lesions can be punctured with a 1.2 mm (18 gauge) needle and this type of needle is also used as a guide when performing fine needle aspiration from solid areas with a 0.6 mm (23 gauge) needle. To obtain material for histological examinations large-bore cutting needle types have been used, but new 20–23 gauge modified Menghini needles have been introduced (Chapter 6). For percutaneous drainage various designs are commercially available. They are mainly based on a modified Seldinger technique or on sheated trocar systems. Puncture of the liver does not include any specific precautions. Neighboring organs should be avoided and patients with a history of hemorrhagic diathesis are excluded.

Cystic focal lesions

The sonographic criteria of a cystic lesion will in most instances enable a true distinction to be made between cystic and solid pathology. Equivocal findings will indicate verification by puncture. It is known that, e.g., sarcomatous, em-

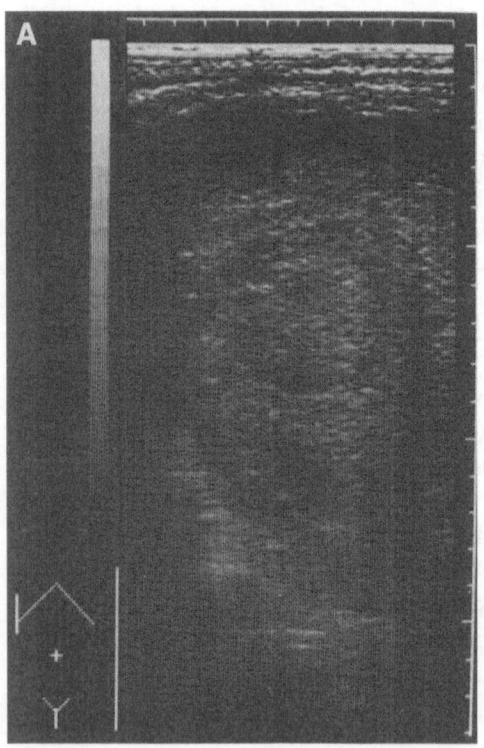 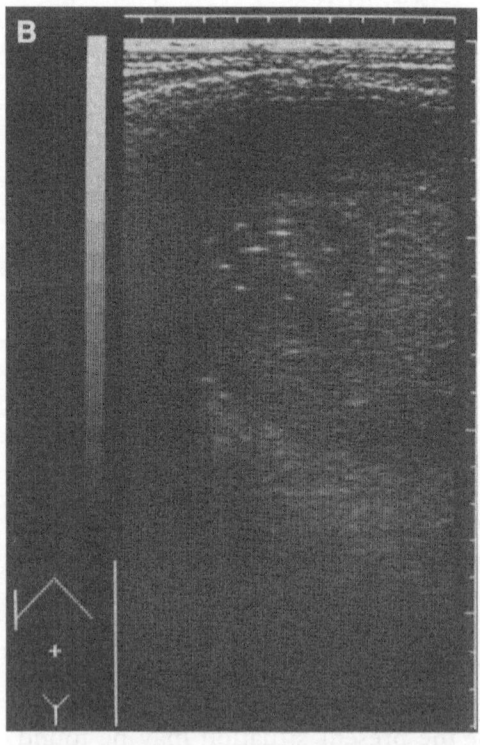

Fig. 7-1. Amebic abscess
A. Longitudinal scan. The echo pattern is only slightly different from that of the surrounding liver tissue. The internal echoes were observed to move with gravity. B. Longitudinal scan. The cavity is emptied through a percutaneously introduced catheter. High level echoes due to air are present in the cavity.

bryonal germinal and melanoma metastases may be quite homogeneous and present with a cystic pattern. On the other hand, many abscesses will contain internal echoes and thereby mimic a solid lesion. Such internal echoes can be seen moving with gravity by changing the position of the patient, but puncture and drainage should still be performed for diagnostic and therapeutic reasons (Fig. 7-1) (Chapter 24).

Simple liver cyst

The presence of simple cysts in the liver is a frequent incidental finding. They may be solitary, sparse, or numerous as seen in patients with hereditary polycystic renal disease. In patients with an unknown infectious focus, puncture is performed to obtain material for culture. If the clinical setting or the sonographic appearance give rise to suspicion of malignant deposit, cytological verification is indicated. Infrequently, the size or location of a cyst may cause symptoms. In our experience repeated punctures have been of little or no value in a number of patients, including patients with polycystic disease, because of recurrence of the cysts. In rare instances a placebo effect has been observed and some patients have developed a kind of "addiction to needling." In one instance drainage by catheter was tried, but the secretory capacity of the 1,000 cc solitary liver cyst caused severe depletion of serum protein. The patient was cured surgically by marsupialization of the cyst to the peritoneal cavity. This seems to be a simple solution in patients with superficially located symptomatic cysts. (Fig. 7-2). No reliable information on shrinkage following the instillation of chemical substances has been reported.

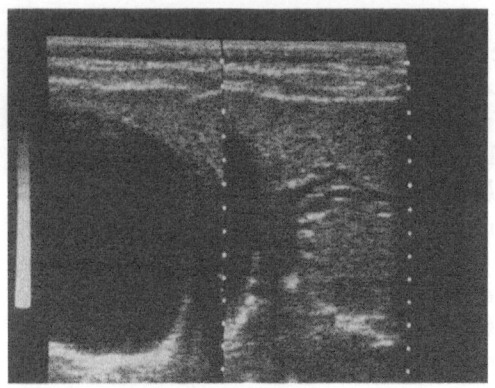

Fig. 7-2. Simple liver cyst
Transverse epigastric scan. The large cyst causes dilation of the biliary branches of the lateral segment of the left liver lobe.

Hydatid liver cyst

The chance that a cystic lesion may be caused by infestation by *Echinococcus granulosus* or *Echinoccocus multilocularis* has been considered a risk at puncture. The sonographic appearance is related to the type of organism involved and is especially influenced by the stage of developement. Following medical treatment with mebendazol we have observed regression. The different sonographic features include thickening of the cyst wall, septation, daughter cysts, internal high level echoes and strongly echogenic walls caused by the presence of calcifications. In spite of this inhomogeneous appearance hydatid disease should be diagnosed in endemic areas, and elsewhere a patient's travel history should be taken prior to puncture of a complex cystic lesion. A few reports have appeared on puncture of hydatid cysts and it seems that the risk of this procedure has been overstated. The diagnostic value of the procedure was demonstrated in a series of 11 hydatid cysts uneventfully punctured (Livraghi et al. 1983). The chlorine and potassium content was found to differ significantly from that obtained from simple cysts. The result is preliminary, but suggests the importance of establishing an awareness of the risk of hydatid cyst puncture. It is well known that the Casoni skin test may be false negative but the serologic complement fixation test is reliable and diagnostic puncture should only be performed when there is no access to this test.

Liver abscess

The frequency and etiology of liver abscesses are influenced by socio-economic conditions and the quality of health care. In the developing countries liver abscesses are common and in many regions the majority will be of amebic etiology. In the highly developed countries the number of imported amebic diseases is increasing and pyogenic liver abscesses seem to occur more frequently in the older age groups. The introduction of sonography has made early detection of liver abscesses possible and the combination with sonographically guided puncture and percutaneous drainage has completely changed the therapeutic approach. As for the *pyogenic liver abscess*, it seems justified to state that surgery should be avoided. The experience of Herlev Hospital (Chapter 24) and our institution in the the treatment of approximately 25 cases of pyogenic liver abscesses indicates a cure rate close to 100% (Fig. 7-3). The-

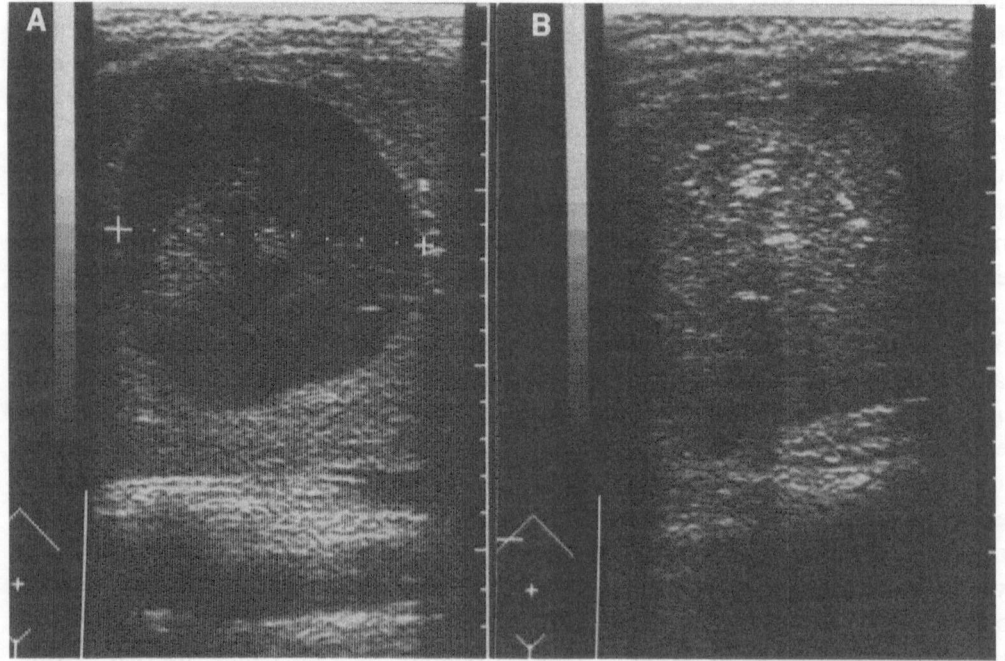

Fig. 7-3. Pyogenic liver abscess
A. Oblique scan. Internal echoes are present. B. Transverse scan. The abscess has been emptied of 1,000 cc of pus through a catheter introduced percutaneously. High level echoes represent bubbles of air.

re were two instances of recurrence after less than 6 months, but the recurrence was also treated percutaneously. In some cases the multiplicity of pyogenic foci prevents drainage, but the diagnostic puncture enables a relevant selection of antibiotic treatment. One case of multiple "micro-foci" caused by *Yersinia enterocolitica* was diagnosed by fine needle aspiration and treated medically (Fig. 7-4). *Amebic abscesses* are treated effectively with metronidazole and chlorokin and puncture is in most instances superfluous if not performed for diagnostic purposes. The content has a characteristic anchovy-sauce color, is odorless and can be examined for *Entamoeba histolytica*. When the medical treatment is terminated the cavity in the liver should no longer be considered an abscess but rather a hepatic necrosis. The different sonographic patterns of resolution have been described in a longitudinal study by Ralls et al.

Hepatic trauma

In most cases the clinical presentation will obviously explain the sonographic findings in hepatic trauma and there will be no need for puncture. In the acute phase fluid can be found in the peritoneal cavity or located subcapsularly. Contusion and hemorrhage in the parenchyma give rise to increased echogenicity. We have observed a similar appearance in a case of parenchymal bleeding due to Goodpasture's syndrome. The echo pattern returned to normal in 10 days. During the resolution of a hematoma the content will become liquified and cystic. In case of symptoms related to a space-occupying effect it may be emptied. In one instance we have observed the resolution ending up with a calcified scar (Fig. 7-5).

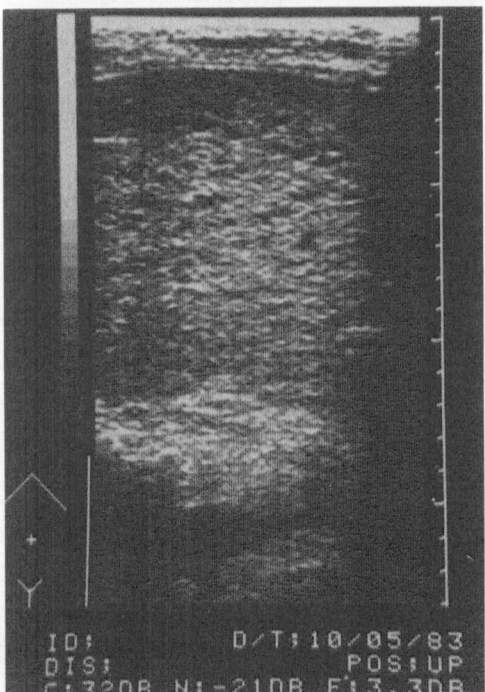

Fig. 7-4. Pyogenic liver "micro-abscesses"
Oblique scan. Multiple pyogenic "micro-cysts" caused by *Yersenia enterocolitica*.

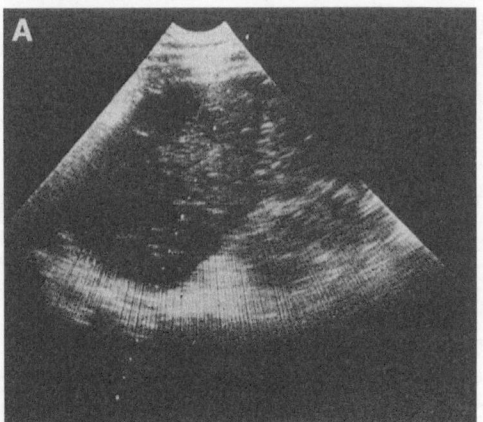

Fig. 7-5. Hepatic trauma
A. Oblique scan 17.11.82. High level echoes in the parenchyma at the dotted line caused by bleeding.

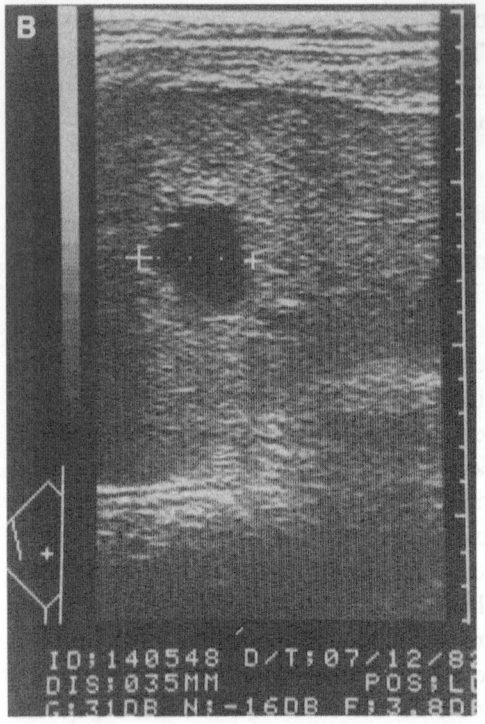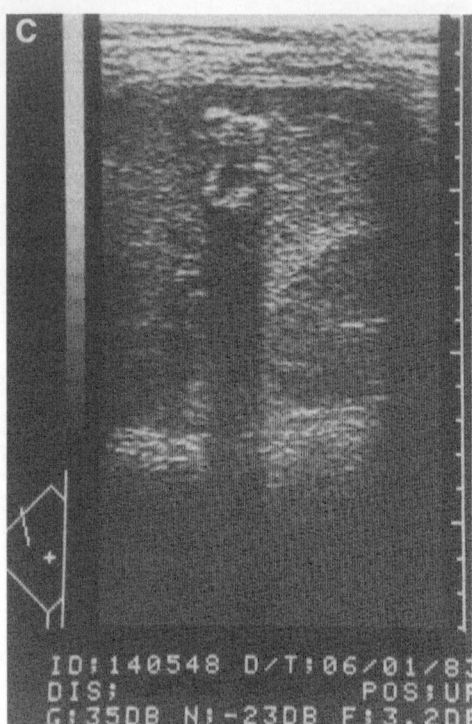

B. Oblique scan 07.12.82. Damaged parenchyma replaced by a cyst. C. Oblique scan 06.1.83. Traumatic cyst transferred to a calcified scar.

Solid focal lesions

In the majority of cases a solid liver lesion will represent a metastatic deposit. However, a number of benign conditions may be detected by sonography. The true distinction in most patients will be of paramount importance for therapeutic and prognostic reasons. *Hemangioma* is the most frequent benign solid lesion. The sonographic appearance has been described in a large number of publications. The diagnosis can be confirmed by a correlative imaging study such as angiography or CT, but fine needle puncture is an alternative. In a series of 24 patients studied by Solbiati et al. (1983) the diagnosis was based on the absence of malignant cells and the presence of endothelial cells in aspirates from fine needle puncture. In 2 cases, transient bleeding occurred. In our experience bleeding can be avoided by performing the puncture through normal liver tissue. Most often hemangiomas are small and characteristically located in the dome of the right lobe of the liver (Fig. 7-6). The correlative imaging procedures should especially be considered in larger lesions with a cystic or a complex appearance when the result of fine needle aspiration does not solve the differential diagnostic problem. The literature dealing with other benign conditions such as *adenoma, focal nodular hyperplasia* and *focal fatty infiltration*

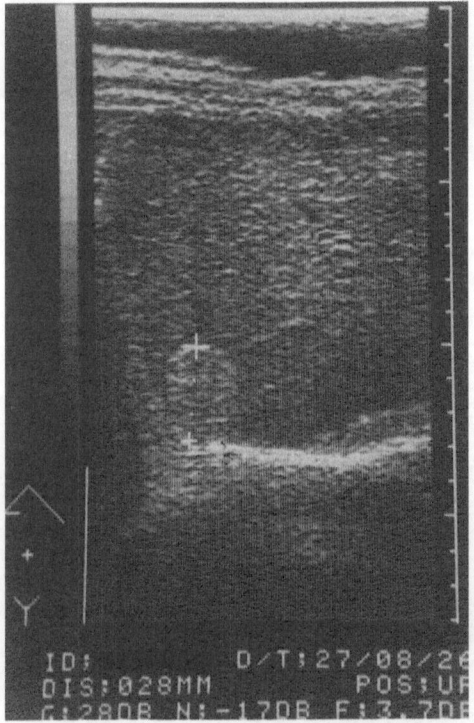

Fig. 7-6. Hemangioma
Transverse scan. A 28 mm echogenic spot localized high in the right lobe of the liver verified as a hemangioma.

is more sparse. The lesions may differ only slightly from the normal parenchyma on sonography but additional features such as change in contour and size may be helpful. Scintigraphy, angiography and CT are of value in these rare conditions and fine needle aspiration has been reported in a few cases (Braun & Dormeyer). Finally, it is commonly recognized that some diffuse liver diseases may cause a focal appearance on sonography. This may represent changes caused by cirrhosis, granulomatous infiltration, chronic congestion and a number of other conditions. Thus, such patients should be expected to constitute the proportion of true negative results in a population aspirated because of sus-

picion of focal liver pathology. Many attempts have been made to correlate the echo pattern of hepatic metastases with the type of primary tumor (Fig. 7-7). So far these investigations have met with only moderate success. The echo pattern of hepatocellular carcinoma deserves special mention as this neoplasm is correlated with cirrhosis, but no prevailing pattern has been confidently proposed (Fig. 7-8).

Results

We have evaluated the accuracy and the diagnostic impact of routinely performed sonographically guided fine needle aspiration from solid focal liver lesions. The data are based on a retrospective review of the results from the Herlev Hospital since 1976 and the Aarhus Kommunehospital since 1982. Altogether 380 patients were examined. The final diagnosis was established by autopsy, operation, a correlative imaging study or the clinical course. The results of the cytological evaluation as compared to the final diagnosis appear in Table 7-1. The predictive value of a malignant cytological diagnosis was 100% (the 95% confidence limits 98–100). Thus the positive test seems highly reliable, which is in accord with other reports as is the absence of false positive diagnoses. The predictive value of a non-malignant cytological diagnosis was 65% (the 95% confidence limits 54–76). These figures indicate that non-malignant cytology is less reliable but is of course influenced by the proportion of patients with non-malignant disease in the investigated population.

The subgroup of patients examined at the Aarhus Kommunehospital com-

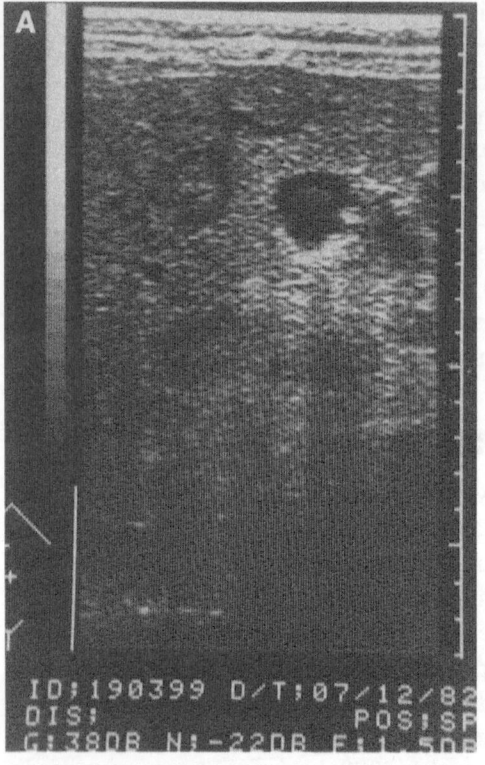 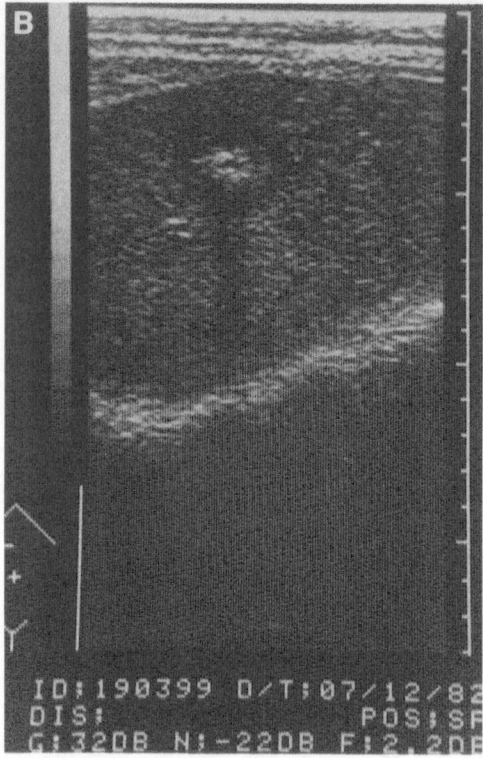

Fig. 7-7. Liver metastases
Two transverse scans. Different sonographic patterns of metastases from an adenocarcinoma of the colon in the same patient.

Table 7-1. Fine needle aspiration from solid focal liver lesions in 380 patients. Correlation between the final diagnosis and the cytological diagnosis

| | | Cytological diagnosis | | |
		Malignant cells	Non-malignant cells	Total
Final diagnosis	Malignant lesion	299	28	327
	Benign condition	0	53	53
		299	81	380

prised 135 patients. The results of this part of the material are presented in Table 7-2. The diagnostic specificity and sensitivity were almost similar. The locations of the primary tumor in the 107 patients with proven liver metastases appear in Table 7-3. False negative results occurred in 3 of 4 cholangiocarcinomas. This may be caused by difficulties in the interpretation of the cytological material. Furthermore, these tumors growing along the biliary tree may be difficult to delineate sonographically and thereby to hit at the aspiration. Metastases originating from the ovary are often small and superficially located which can explain the false negative result in 2 instances and the diagnosis of liposarcoma presents definite cytological problems as experienced in 1 case. No explanation can be given for the last 5 false negative results.

Table 7-2. Fine needle aspiration from solid focal liver lesions in 135 patients. Correlation between the final diagnosis and the cytological diagnosis

		Cytological diagnosis		
		Malignant cells	Non-malignant cells	Total
Final diagnosis	Malignant lesion	96	11	107
	Benign disease	0	28	28
		96	39	135

Diagnostic specificity: 96/96 = 100% (95% confidence limits: 96–100)
Diagnostic sensitivity: 28/39 = 72% (95% confidence limits: 55–85)

In Table 7-4 the final diagnoses are listed in the patients with a true negative cytological diagnosis. Some of these patients turned out to suffer from a disease related to the hepatobiliary system which may explain an echo pattern arousing suspicion of focal pathology.

Table 7-3. Location of primary tumor in 107 patients with malignant involvement of the liver

Lung	20
Liver	4
Gall bladder	3
Biliary tree	4
Pancreas	19
GI-tract	20
Mamma	7
Ovary	3
Testis	1
Kidney	2
Urinary bladder	3
Skin	2
Unknown	19
	107

Table 7-4. Final diagnosis in 28 patients with a true non-malignant cytological diagnosis

Malignant primary, no liver metastases	9
Jaundice/elevated alk. phosphatase	5
Gall bladder disease	6
Cirrhosis	5
Amyloidosis	1
Hemangioma	2
Total	28

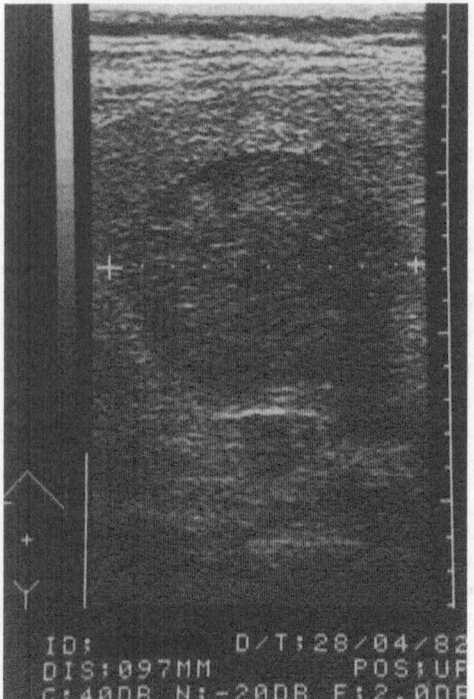

Fig. 7-8. Hepatocellular carcinoma
Transverse scan. A 97 mm slightly hypoechoic hepatoma in an otherwise normal liver. Clinically the disease presented with symptoms caused by medullary compression because of metastasis to the spine.

In others the presence of jaundice, elevated alcaline phosphatase or a known malignant primary would raise the threshold of performing aspiration from suspicious areas. The rest of the patients had definite, but benign focal lesions or the focal appearance of a cirrhotic liver. According to the reports of the sonographic examination the diagnosis was definite in 96 instances and suspected in 39. By also using the fine needle aspir-

51

ation the number of true and definite diagnoses was increased by 32 to a total of 128, or 94%.

Hazards of puncture are dealt with elsewhere (Chapter 26), but it should be mentioned that 1 case of a cutaneous implantation metastasis at the puncture site was recognized in this series. The procedure was performed by an inexperienced examiner and the 1.2 mm guide needle was unintentionally introduced into a superficially located lesion in a patient with advanced hepatocellular carcinoma. Thus, the tumor was not only punctured with a fine needle. This risk can be minimized by the use of a stop screw on the guide needle when training sonographers in the technique of fine needle aspiration. No other complications were encountered among the 135 patients examined.

We find it fair to conclude that fine needle aspiration from focal liver lesions increases the accuracy of hepatic sonography, reduces the number of diagnostic procedures and especially thereby facilitates patient management.

References

Christensen M, Rødbro P. The diagnostic value of liver scintigraphy to disclose metastases in patients with suspected or proven gastrointestinal cancer. *Dan Med Bull* 1982; 29: 206.

Haubek A, Gammelgaard J, Grønvall S, Holm H H. Ultrasonically guided percutaneous puncture and biopsy techniques. In: Wilkins R A, Viamonte Jr M, eds. Interventional Radiology. Oxford: Blackwell Scientific Publications, 1982: 373.

Hadidi A. Ultrasound findings in liver hydatid cysts. *J Clin Ultrasound* 1979; 7: 365.

Gharbi H A, Hassine W, Brauner M W, Dupuch K. Ultrasound examination of the hydatic liver. *Radiology* 1981; 139: 459.

Livraghi T, Bosoni A, Giordano F. Diagnosis of hydatid and non-parasitic abdominal cysts by percutaneous aspiration – Value of physiochemical examination. Third International Congres on Interventional Ultrasound, Copenhagen, 1983.

Berry M, Bhargava S, Bazaz R. Liver abscess, sonographic diagnosis and management. Third International Congress on Interventional Ultrasound, Copenhagen, 1983.

Grønvall S, Gammelgaard J, Haubek A, Holm H H. Drainage of abdominal abscesses guided by sonography. *Am J Roentgenol* 1982; 138: 527.

Ralls P W, Quinn M F, Boswell W D, Colletti P M, Radin D R, Halls J. Patterns of resolution in successfully treated hepatic amebic abscess: Sonographic evaluation. *Radiology* 1983; 149: 541.

Bree R L, Schwab R E, Neiman H L. Solitary echogenic spot in the liver: Is it diagnostic of a hemangioma? *Am J Roentgenol* 1983; 140: 41.

Solbiati L, Ierace T, Livraghi T, Masciadri N, Vettori C. Fine neddle biopsy of liver hemangioma. Third International Congress on Interventional Ultrasound, Copenhagen, 1983.

Braun B, Dormeyer H H. Ultrasonically guided fine needle aspiration biopsy of hepatic and pancreatic space occupying lesions and percutaneous abscess drainage. *Klin Wochenschr* 1981; 59: 707.

Hillmann B J, Smith E H, Gammelgaard J, Holm H H. Ultrasonic-pathologic correlation of malignant hepatic masses. *Gastrointest Radiol* 1979; 4: 361.

Schwerk W B, Schmitz-Moormann P. Ultrasonically guided fine needle biopsies in neoplastic liver diseases: Cytohistologic diagnoses and echo pattern of lesions. *Cancer* 1981; 48: 1469.

CHAPTER 8

Ultrasonically guided cholangiography and bile drainage

Masatoshi Makuuchi & Yasutsugu Bandai

Large intrahepatic ductal structures such as the hepatic veins and the portal venous branches are clearly seen with ultrasound. The normal intrahepatic bile ducts are too small to be demonstrated except at the porta hepatis. However, if the intrahepatic bile ducts are dilated due to obstruction, it can be clearly seen with ultrasound. Goldberg introduced the term, "Ultrasonic Cholangiography" indicating the clearness of the bile duct on the sonogram.

The ultrasonically guided puncture technique has been used for many purposes. However, the techniques was not applied to the biliary system until we reported the ultrasonically guided percutaneous transhepatic cholangiography (UG-PTC) procedure. First, experimental studies with swine were performed using a manual compound system; subsequently UG-PTC was carried out in clinical cases in 1976. With the real-time system later used, the needle tip and guide wire were clearly seen. With this technique ultrasonically guided percutaneous transhepatic bile drainage (UG-PTBD) and ultrasonically guided percutaneous transhepatic gall bladder drainage (UG-PTGBD) have been performed since March and July 1979, respectively.

Ultrasonically guided-percutaneous transhepatic cholangiography (UG-PTC)

UG-PTC has been carried out since October 1976. The point for insertion of the needle was freely chosen on the basis of the most dilated bile duct closest to the body surface. Moreover, the site of puncture was not influenced by the lungs and the ribs. The route for the conventional PTC under fluoroscopy was not used. In patients with all parts of the intrahepatic bile ducts dilated, the peripheral end of the left hepatic duct was selected for puncture from the epigastrium. In patients with dilation of only part of the intrahepatic bile ducts, the dilated bile duct was punctured closest to the body surface, avoiding the gall bladder.

After sterilization of the skin, the puncture transducer was held at the preselected point from which the dilated bile duct was best seen. With the patient holding his breath, the needle was inserted. An outer guide needle was used to penetrate the skin and fascia to prevent the needle bending. After complete aspiration of bile, a radiopaque dye was injected.

UG-PTC was performed successfully

and without complications on 51 occasions in 47 patients. Patients' diseases are listed in Table 8-1. In UG-PTC, the

Table 8-1. UG-PTC cases (1976–1983)

Disease	Cases	Occasions
Ca. of Head of Pancreas	7	7
Ca. at Porta Hepatis	10	11
Ca. of Main Bile Duct	1	2
Ca. of Gallbladder	3	4
Ca. of Duodenal Papilla	2	2
Choledocholithiasis	7	8
Hepatolithiasis	9	9
Stenosis of Hepatojejuno-stomy	3	3
Others	5	5
Total	47	51

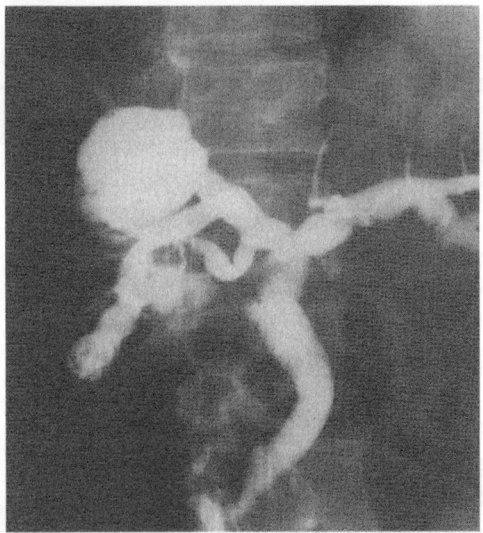

Fig. 8-2. UG-PTC for hepatolithiasis
The distal end of the postero-inferior area duct is punctured. Many stones are impacted in the postero-inferior area duct and the postero-superior area duct is changed into a large cyst containing several stones. If the distal portion of the bifurcation of the right posterior duct had not been punctured in this patient, the postero-superior and the postero-inferior area ducts could not have been opacified.

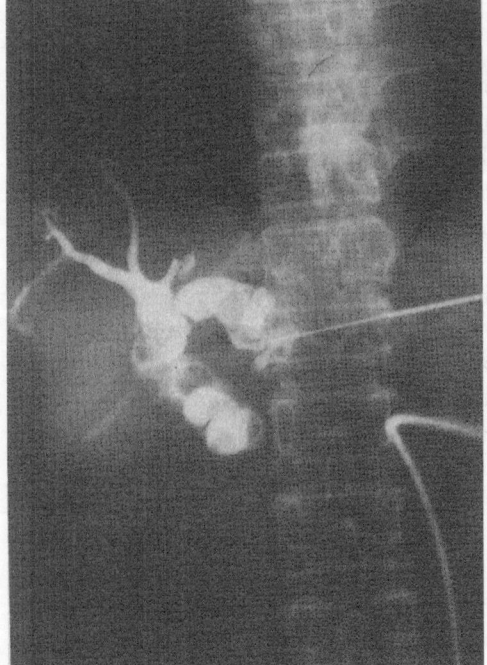

Fig. 8-1. UG-PTC for hepatolithiasis
The left hepatic duct is punctured distal to the stricture. Several stones are demonstrated in the left hepatic duct and the common bile duct. If the conventional PTC from the right intercostal space had been performed in this patient, the left bile duct distal to the stricture could not have been opacified and several more punctures would have had to be made.

puncture site was mainly from the epigastrium to the left hepatic duct system (Fig. 8-1). The right hepatic duct system was punctured in patients with obstruction at the porta hepatis or with right hepatic duct obstruction due to hepatolithiasis (Fig. 8-2). Any duct could be punctured selectively. In patients with dilation of only part of the intrahepatic bile ducts because of stricture, the part peripheral to the stricture could be punctured and the whole biliary tract demonstrated by one opacification (Fig. 8-1).

Since Arner et al. introduced fluoroscopy for PTC, it has generally been used for differentiation of jaundice. However, there are some disadvantages:

55

1) The puncture point and the direction of the needle are determined only from knowledge of the normal anatomy of the bile duct system. 2) Often more than one puncture is necessary. 3) If the needle is not inserted directly into the bile duct, cholangiograms obtained are sometimes obscured by the leakage of contrast medium. 4) Selective bile duct puncture is very difficult, for example the left hepatic duct puncture by Glenn's right lateral approach. The conventional PTC method is a blind puncture followed by X-ray control, whereas the UG-PTC permits puncture of the bile ducts under simultaneous ultrasonic visualization.

After the UG-PTBD procedure was introduced, UG-PTC was not performed as preparation for bile drainage but was reserved for patients with partial dilation of the intrahepatic bile duct.

Ultrasonically guided-percutaneous transhepatic bile drainage (UG-PTBD)

UG-PTBD has been carried out from March 1979 in 183 patients with dilated bile ducts demonstrated by ultrasound. Diseases of the patients are listed in Table 8-2. A real-time sector scanner was used. The needle for PTBD was 17 gauge, 20 cm in length with a curved tip. The guide wires first used had a 'J' tip (3 mm) and a straight tip, 0.813 mm in diameter and 120 cm long. Then a guide wire with a movable core, J tip (6 mm), 0.9 mm in diameter and 80 cm long was used. Dilators from 6 to 8 French were used for dilating the puncture site of the bile duct. The drainage tube was made of 7 or 8 French polyethylene tubing.

After the entire biliary system has

Table 8-2. UG-PTBD cases (1979–1983)

Disease	Cases	No. of Intubations Left	Right
Primary Ca. of Liver	6	7	2
Ca. of Bile duct at Porta Hepatis	32	31	15
Gallbladder Carcinoma	20	21	5
Ca. of Main Bile Duct	16	14	3
Ca. of Head of Pancreas	41	46	2
Ca. of Duodenal Papilla	4	4	0
Hepatolithiasis	6	4	3
Choledocholithiasis	15	12	2
Inflammatory Mass of Pancreas	5	1	4
Benign Stricture	4	5	2
Recurrence or Metastasis of Ca.	30	25	8
Others	4	4	0
		174	46
total	183	220	

been examined by ultrasound, the puncture site of the bile duct is selected. We chose to insert the needle from the epigastrium into the inferior duct of the left lateral lobe in patients with dilation of the whole biliary system because the left side is more readily delineated by ultrasound and its peripheral bile ducts are easier to puncture than those of the right side. However, if the left lobe was invaded by tumor or if carcinoma of the porta hepatis had invaded the left side much more than the right side, the right hepatic duct was punctured. The bile duct was preferably about 6 mm or more in diameter and as peripheral as possible at the point of entry.

After adjusting the scanner to the position at which the puncture line on the oscilloscope intersects the target duct (Fig. 8-3A), an incision is made in the skin and then the abdominal wall is penetrated with an outer guide needle. Through this the PTBD needle is introduced into the bile duct while the patient holds his breath (Fig. 8-3B). Following removal of the stylet, correct positioning of the needle tip is confirmed by the spontaneous drainage or aspiration of bile and the spring guide wire is inserted under ultrasonic control (Fig. 8-3C). After the guide wire has been smoothly inserted, the scanner is detached from the PTBD needle, and the course and position of the guide wire is seen by real-time sonography (Fig. 8-3D). When the guide wire has passed far enough into the bile duct, the PTBD needle and the outer guide needle are withdrawn, leaving the wire behind. Then, the puncture in the bile duct is enlarged with dilators, and the drainage tube is advanced as far over the wire as possible. Advancing of the dilators and the drainage tube is monitored by ultrasound and/or fluoroscopy (Fig. 8-3E, 8-3F). Following insertion of the drainage tube, the wire is removed and as much bile as possible aspirated. Cholangio-

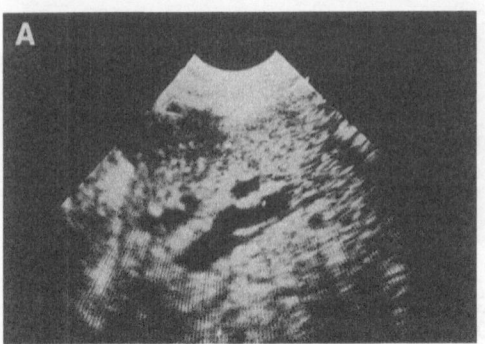

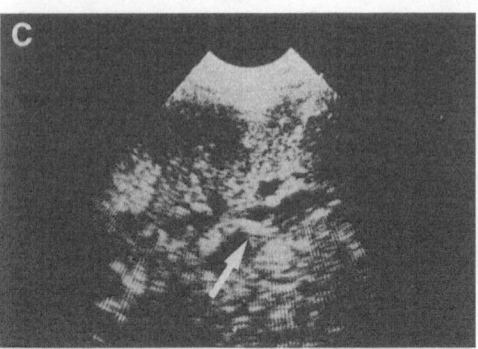

Fig. 8-3. Sonogram of UG-PTBD procedure
A. The dilated right postero-inferior area duct is clearly demonstrated and the scanner is adjusted until the puncture line transects the target.

C. The spring guide wire is inserted through the PTBD needle and the tip of the guide wire (arrowhead) appears in the bile duct. The spring guide wire reflects very strong echoes with reverberation artifacts behind.

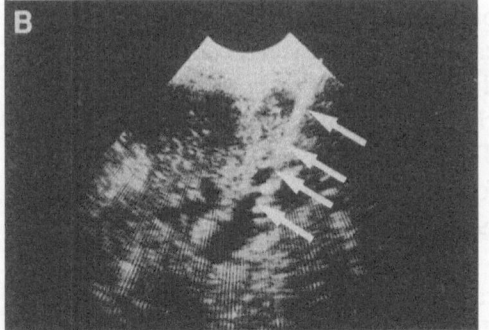

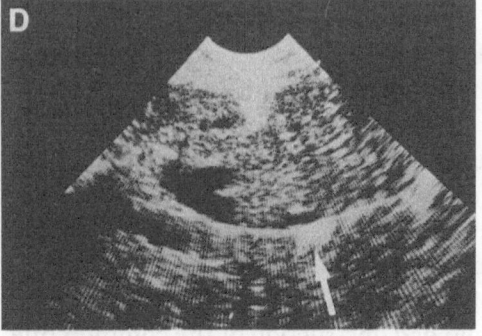

B. The PTBD needle is inserted into the bile duct. The whole length of the needle (arrows) is clearly seen as a line of strong echoes.

D. The guide wire has been inserted far enough into the bile duct. This sonogram is taken after the scanner is detached from the needle. Arrow indicates the comet tail-like reverberation artifact which is characteristic of the guide wire.

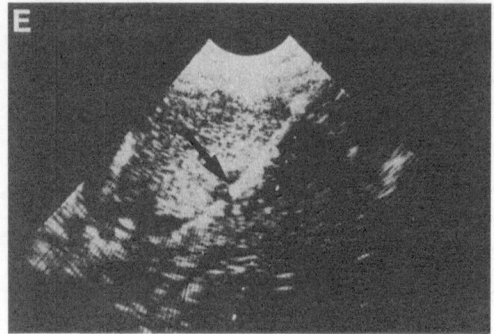

E. This sonogram shows the catheter advanced over the guide wire. Arrow indicates the tip of the catheter. The catheter does not produce the comet tail-like echos.

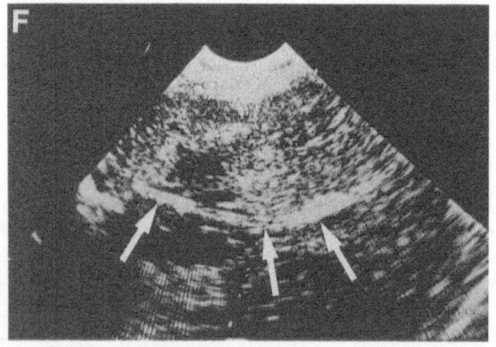

F. This sonogram shows the catheter after the spring guide wire is withdrawn. The catheter (arrows) is demonstrated as strong parallel lines.

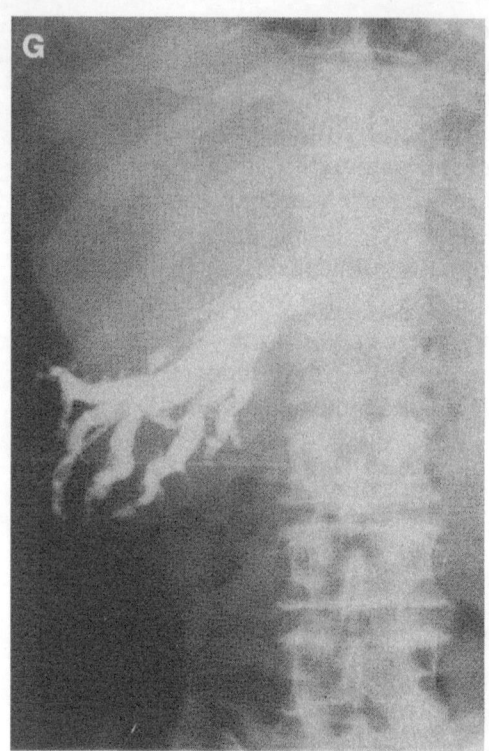

G. Completion cholangiogram of this patient. Only the right postero-inferior area duct is opacified.

graphy is then performed via the tube to confirm that the tube is in the correct position (Fig. 8-3G). Since March 1981 a silastic balloon catheter has been used (Fig. 8-4). In several patients, internal drainage has been attempted.

UG-PTBD has been performed 224 times in 183 patients. In 4 patients, the guide wire was introduced in the bile duct but the 7F dilator and the drainage tube did not penetrate the anterior wall of the bile duct. Two of these patients, who had obstruction of the common bile duct, underwent gall bladder drainage. The 3rd patient, who had carcinoma in the porta hepatis, had a right-side hepatic duct drained. In the last patient, a catheter was introduced along the bile duct but bile could not be aspirated. By cholangiography, stones in the common bile duct were demonstrated and the catheter was placed outside the bile duct. Cholecystectomy and removal of choledocholithiasis was performed. Thus, UG-PTBD was successfully performed in 220 out of 224 attempts (98.2%).

In 10 patients with 11 intubations the catheter penetrated the posterior wall of the bile duct. This caused no complications, the bile drainage was sufficient and jaundice subsided because the catheter has 3 lateral holes and these were ajusted to the bile duct. To prevent pen-

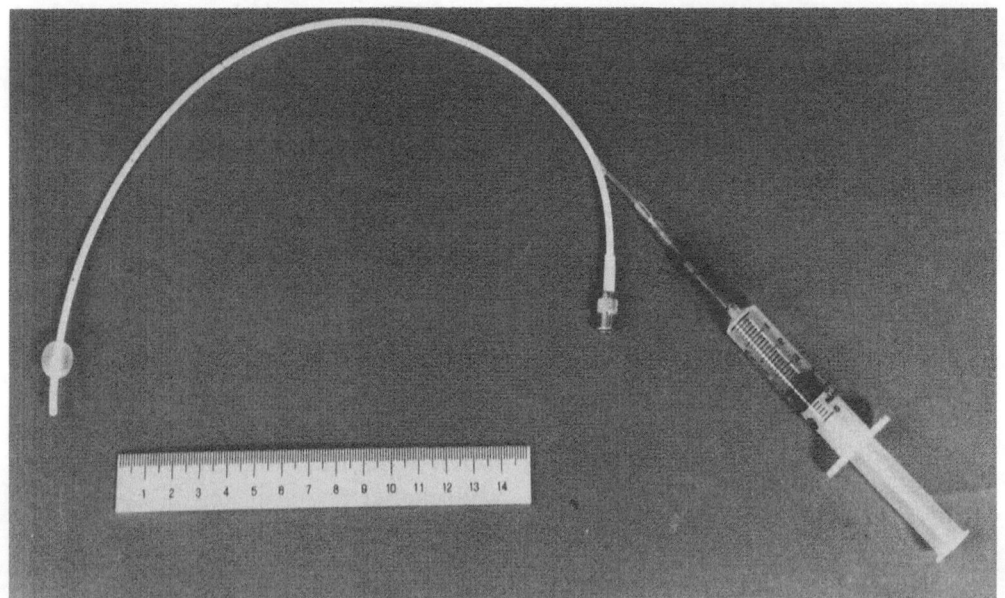

Fig. 8-4. Silastic balloon catheter
Not only the tube but also the balloon is made of silicone. Any size of catheter is available from 6F.

etration of the posterior wall of the bile duct, the guide wire with core was never advanced when more than slight resistance was felt. In such situations the needle was pulled out 3 to 5 mm. At the same time, the scanner was tilted to an optimum position, i.e. a narrow angle of the needle to the bile duct. Once the guide wire with the core had penetrated the posterior wall of the bile duct, it was virtually impossible to insert the catheter. These incomplete intubations occurred 11 times out of 220 intubations (5%). The main complication after drainage was that the catheter slipped out from the bile duct. This happened with 27 catheters in 23 patients (12.3%). Sixteen catheters were replaced by UG-PTBD.

The drainage tube slipped out from 2 to 47 days after intubation. The causes were not always clear but most of the patients had either a severe cough, an attack of asthma or intermittent positive pressure breathing for preoperative care or endoscopic examination.

To prevent the catheter from slipping out of the bile duct, a 7F and 8F silastic catheter with balloon was designed. In 34 out of 39 attempts, the drainage tubes were replaced with balloon catheters. One tube slipped out (2.9%) due to deflation of the balloon.

With the UG-PTBD procedure, the left lateral inferior bile duct could be drained even in a patient with a large tumor in the right lobe of the liver (Fig. 8-5).

Bleeding after the UG-PTBD procedure occurred in 7 patients (3.8%). Four patients had slight bleeding which stopped soon after the drainage. In 1 patient, the bleeding continued and the bile drainage was inadequate. There-

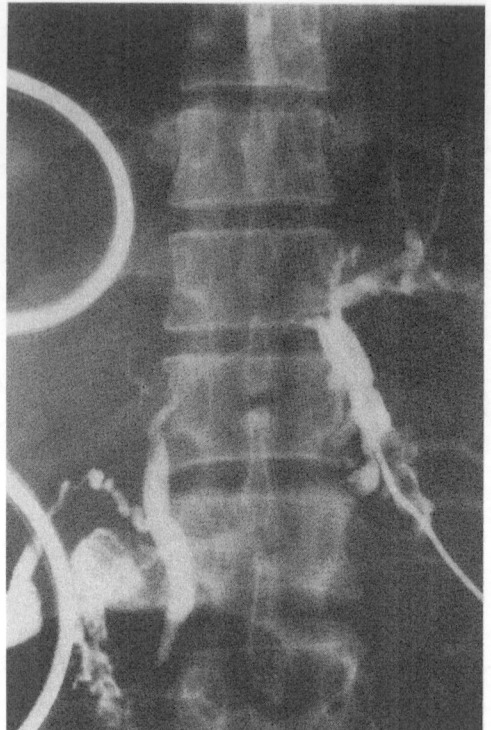

Fig. 8-5. UG-PTBD in a patient with hepatoma and obstructive jaundice
The left lateral-inferior area duct is punctured and drained. The right lobe of the liver and the left medial segment are occupied by a hepatoma which invades the left lateral segment. If the ultrasonically guided puncture technique had not been used, it would have been virtually impossible not only to perform bile drainage but also to perform thin needle cholangiography.

fore, UG-PTBD was performed again and bleeding stopped 2 days after reintubation. In the other 2 patients, bleeding started 5 and 7 days after drainage. In the former patient, bleeding continued for 2 days and then stopped. In the latter patient, bleeding continued for 8 days until he expired due to carcinoma of the bile duct.

Hypotension after drainage was experienced by 5 patients, of whom 3 used hypertensive drugs. They had also cholangitis but recovered 4, 6 and 15 hours later. Among the 183 patients, 19 had suppurative cholangitis. However, endotoxin shock after bile drainage was not experienced (Fig. 8-6).

By conventional PTBD methods, the direction of needle insertion is limited to horizontal or perpendicular. However, the UG-PTBD method is not restricted to these two directions and puncture can be performed at a narrow angle to the bile duct. For this purpose, as well as for visualization of the needle and the guide wire, a sector scanner is much better than an electronic linear array scanner.

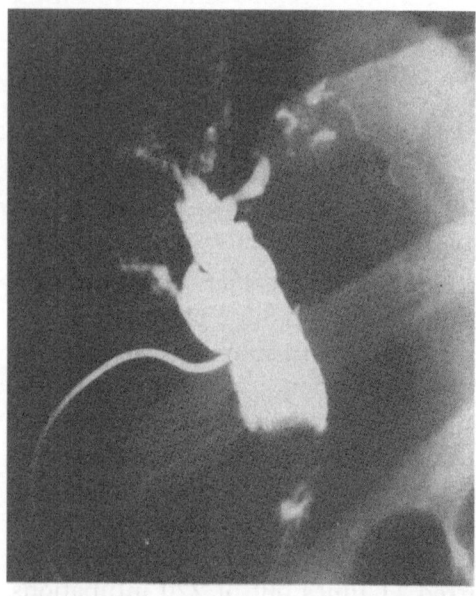

Fig. 8-6. UG-PTBD in a patient with hepatolithiasis and acute suppurative cholangitis
Only the right antero-superior duct is opacified and a large stone is impacted at the porta hepatis. Many small liver abscesses are demonstrated at the periphery of the bile duct. In this case, if conventional PTC is performed, endotoxin shock cannot be avoided. This patient was without jaundice. (Reproduced from: Makuuchi M et al. Ultrasonically guided percutaneous transhepatic bile drainage; A single-step procedure without cholangiography, Radiology 1980; 136; 167.)

The minimum caliber of the selected bile duct for puncture depends on the skill of the operator. The skilful operator can puncture a bile duct 3 or 4 mm in diameter. With conventional PTBD methods, it is very difficult to puncture a peripheral bile duct. (Fig. 8-5).

The main causes of the catheter slipping out from the bile duct were coughing and deep breathing. The catheter slipped out to the free abdominal cavity due to the displacement between the abdominal wall and the liver surface. At first, a loop of the tube is formed between the liver and the abdominal wall and then the tube slips out. Therefore, it is important to minimize the length of the catheter in the abdominal cavity during respiration. For this purpose, puncturing should be carried out at the mid-point of breathing. Puncturing when a patient inspirates deeply should be avoided.

As the UG-PTBD method is a single-step procedure without cholangiography, intraductal pressure is not raised by injection of the contrast medium, so UG-PTBD is safer in patients with suppurative cholangitis (Fig. 8-6). In this series, 5 patients became hypotensive after the drainage procedure but no endotoxin shock was experienced. Takada was the first to perform PTBD in a patient with suppurative cholangitis in Japan and emphasized that PTBD was not contraindicated in cholangitis. However, 4 out of 24 patients with suppurative cholangitis were lost in his series due to endotoxin shock. The UG-PTBD method seems much safer than the conventional two-step procedure.

The left bile duct was punctured and intubated 174 times out of 220 (79%). With the left side approach, the puncture site is not disturbed by the ribs and the large bile duct is closer to the skin srface than on the right side. So more peripheral bile ducts can be punctured with the left side approach (Fig. 8-5). Moreover, the movements of the liver with respiration are less pronounced on the left than on the right side.

Other merits of UG-PTBD are; 1) puncture of the large blood vessels can be avoided because ultrasound can demonstrate these vessels simultaneously. 2) Irradiation can be avoided or minimized. 3) UG-PTBD can be performed in the patient's room if desired.

The disadvantages of UG-PTBD are;
1) Training of doctors in real-time scanning is necessary.
2) An appropriate real-time system with puncture attachment must be available.

Ultrasonically guided-percutaneous transhepatic gall Bladder drainage (UG-PTGBD)

UG-PTGBD has been carried out in 36 patients with a dilated gall bladder from July 1979 to July 1983. Diseases of the patients are listed in Table 8-3. The gall bladder was punctured from the 7th or 8th intercostal spaces or from the right

Table 8-3. UG-PTGBD cases (1980–1983)

Disease	No. of Cases
Cholecystitis	16
Choledocholithiasis	7
Ca. of Common Bile Duct	2
Chronic Pancreatitis	2
Ca. of Head of Pancreas with Pan-liver Metastasis	3
Recurrence of Gastric Ca.	4
Primary Sclerosing Cholangitis	1
total	36

subcostal area through the liver parenchyma. In the first case, a teflon sheathed needle was used and the gall bladder was drained by the sheath. In 18 patients, UG-PTGBD was performed by the same procedure as UG-PTBD and a straight or a pigtail catheter was introduced in the gall bladder. In 6 patients, a J-shaped catheter with a hard metal stylet was directly inserted in the gall bladder under ultrasonic guidance. As a fourth alternative, in 11 patients a 14 gauge teflon sheathed needle was used and a silastic balloon catheter was introduced through the teflon sheath to the gall bladder, whereupon the balloon was inflated.

All these 36 UG-PTGBD procedures were successfully performed. Puncture of the gall bladder was much easier than puncture of the intrahepatic bile ducts due to its large size. In patients with carcinoma of the head of the pancreas and pan-liver metastases, stable bile drainage could be continued until they expired due to carcinoma (Fig. 8-7). In patients with acute cholecystitis, pure bile could be obtained from the tube several days after drainage and the cystic duct was recanalized (Fig. 8-8).

The main complication after UG-PTGBD was bleeding in 4 patients (11.1%). In 1 patient bleeding occurred only during the UG-PTGBD procedure and it stopped just after drainage was accomplished. In 2, bleeding continued for several hours after drainage. In the last case, the catheter was obstructed a few times due to a blood clot. However, blood transfusion was not necessary in any of the 4 patients. Slipping out of the catheter, hypotension after drainage or cholascos was not experienced in this series.

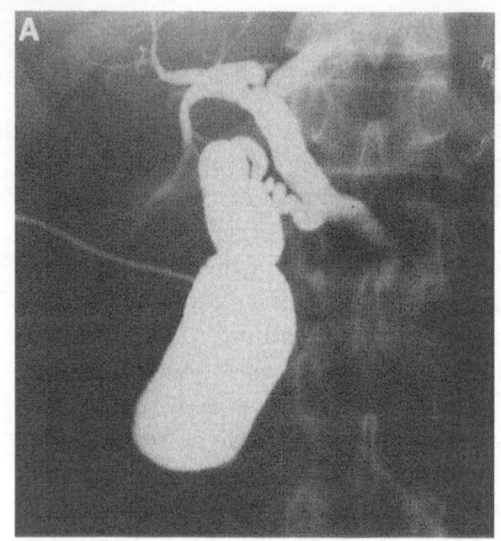

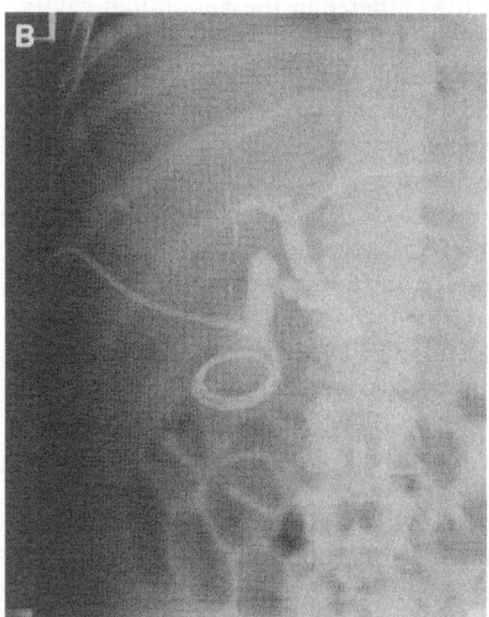

Fig. 8-7. UG-PTGBD in a patient with carcinoma of the head of the pancreas

A. The whole liver is invaded by metastases and the intrahepatic bile duct is only slightly dilated. The bile duct is irregular due to tumors, and the right anterior bile duct is not opacified.

B. In this case, a pig-tail catheter with double loops is introduced in the gall bladder to prevent it from slipping out.

Gall bladder drainage is useful not only in patients with common bile duct obstruction but also in acute cholecystitis. If intravesical pressure becomes negative due to a bile drainage procedure, bile leakage will never occur. In patients with suppurative cholecystitis and in poor risk patients, UG-PTGBD is useful. Using UG-PTGBD, the patient's condition improves and recanalization to the main bile duct happens in almost all patients. So the total biliary system can be visualized by injection of contrast material into the drainage tube (Fig. 8-8). In patients with postopera-

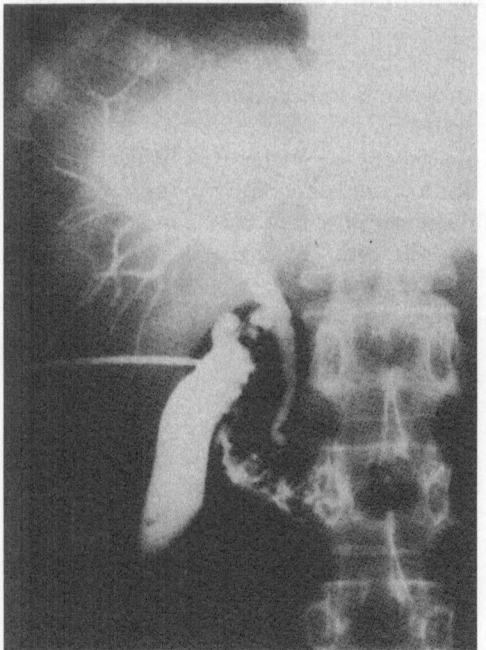

Fig. 8-8. UG-PTGBD in a patient with postoperative acute cholecystitis
Cholecystography is performed several days after UG-PTGBD. The cystic duct is recanalized and the whole biliary system is opacified.

tive acute cholecystitis, the gall bladder can be preserved.

Change in indication of PTC

Before ultrasound was introduced for differentiation of jaundice, PTC was performed to differentiate obstructive jaundice. Now, jaundice is easily differentiated by ultrasound. Therefore, this indication for PTC has been taken over by ultrasound. Another major indication for PTC was preparation for PTBD. However, as the UG-PTBD method is a single-step procedure without cholangiography, PTC is not indicated for preparation of bile drainage either. If the patient is suspected of having bile duct diseases, we first perform ultrasound examination to determine the degree and localization of bile duct dilation. If the patient has dilated bile ducts in the entire liver, UG-PTBD is indicated. If the patient has extrahepatic bile duct dilation, endoscopic retrograde cholangiography is indicated. If the patient has localized dilation of the intrahepatic bile ducts, for example congenital cystic dilation or hepatolithiasis, UG-PTC is indicated. However, if the patient has cholangitis, UG-PTBD is indicated. Therefore, indications of PTC are very few. Actually, since 1979, UG-PTBD has been performed in 183 patients and UG-PTC in only 20 patients. If the UG-PTBD method had not been introduced, almost all 183 patients would have had PTC performed.

Most large institutions in Japan which previously performed PTC and PTBD now use the ultrasonically guided puncture technique.

References

Goldberg B B. Ultrasonic cholangiography. *Radiology* 1976; 118: 401.

Holm H H, Kristensen J K (eds). Ultrasonically guided puncture technique. Copenhagen: Munksgaard and Baltimore: University Park Press, 1980.

Makuuchi M, Kamiya K, Beppu T et al. Percutaneous transhepatic cholangiography under ultrasonic guidance. *Acta Hepatol Jpn* 1977; 18: 435, (in Jpn).

Makuuchi M, Beppu T, Kamiya K et al. Echo guided percutaneous transhepatic cholangiography with puncture transducer. *Jpn J Surg* 1978; 8: 165.

Makuuchi M, Bandai Y, Ito Y et al. Ultrasonically guided percutaneous transhepatic bile drainage; A single-step procedure without cholangiography. *Radiology* 1980; 136: 165.

Saitoh M, Watanabe H, Ohe H, Tanaka S, Itakura Y, Date S. Ultrasonic real-time guidance for percutaneous puncture. *J Clin Ultrasound* 1979; 7: 269.

Arner O, Hagberg S, Seldinger S I. Percutaneous transhepatic cholangiography. *Surgery* 1962; 52: 561.

Glenn F, Evans J A, Mujahed Z, Thorbjarnarson B. Percutaneous transhepatic cholangiography. *Ann Surg* 1962; 165: 451.

Ohto M, Ono T, Tsuchiya Y, Saisho H. Cholangiography and pancreatography. Tokyo: Igaku-Shoin, 1978.

Takada T, Hanuu F, Kobayashi S, Uchida Y. Percutaneous transhepatic cholangical drainage: direct approach under fluoroscopic control. *J Surg Oncol* 1976; 8. 83.

Takada T. Illustrated manual of percutaneous transhepatic cholangiography with drainage. Tokyo: Igaku-Shoin, 1978, (in Jpn.).

Goldstein L I, Sample W F, Kadell B M, Weiner M. Gray scale ultrasonography and thin-needle cholangiography. *JAMA* 1977; 238: 1041.

Conrad M R, Landay M J, Janes J O. Sonographic "parallel channel" sign of biliary tree enlargement in mild to moderate obstructive jaundice. *Am J Roentgenol* 1978; 130: 279.

Koenigsberg M, Wiener S N, Walzer A. The accuracy of sonography in the differential diagnosis of obstructive jaundice: A comparison with cholangiography. *Radiology* 1979; 133: 157.

CHAPTER 9

Intraoperative puncture of the liver guided by ultrasound

Masatoshi Makuuchi, Hiroshi Hasegawa & Susumu Yamazaki

Intraoperative ultrasonic examination is now commonly used in many fields of surgery, i.e. neurosurgery, heart, hepatic, biliary, pancreatic, urologic and vascular surgery. In the field of hepatic surgery, intraoperative ultrasonic examination is especially useful, because the ductal structures and small tumors in the liver are neither visible nor palpable from its surface. With ultrasound, these structures are visible and even more clearly seen with intraoperative than with external scanning. Furthermore, if ultrasonically guided puncture is carried out intraoperatively, the operative procedure itself becomes simpler and a more accurate radical resection of the liver can be performed. In our institute, intraoperative ultrasonic examination and ultrasonically guided puncture are now indispensable for hepatectomy, especially in the small hepatocellular carcinoma associated with liver cirrhosis.

Development of intraoperative ultrasonic probes

Since 1976, we have performed intraoperative ultrasonic examination with a conventional manual compound scanning system, a mechanical sector scanning system and a linear array electronic scanning system. These scanners, however, are not suitable for intraabdominal scanning because scanning is disturbed by the costal arch and the abdominal wall. For intraoperative ultrasonic examination, the ultrasonic real-time system must be small, have fine resolution, little electronic noise in the near field, and be electrically safe. For this purpose we devised the first ultrasonic probe for intraoperative ultrasonic examination in October 1979.

Since then, several improvements have been added to the probe and a puncture adaptor has also been devised. Fig. 9-1A shows the present probe with the puncture adaptor. This probe, which is commercially available, has 48 5 MHz crystals, it is 9 cm long and allows an ultrasonically visualized field of 6 cm width. To achieve better resolution, another probe of 7.5 MHz frequency has been developed (Fig. 9-1B). This probe has better resolution but sound penetration into the liver parenchyma is not as effective. For the visualization of all parts of the liver, the 5 MHz transducer is useful and the 7.5 MHz transducer is suitable for more precise evaluation of a small lesion detected by the 5 MHz transducer.

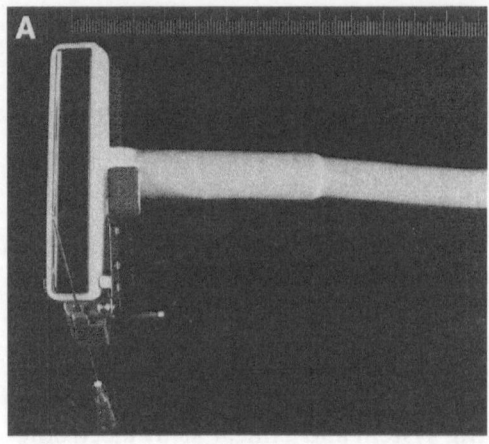
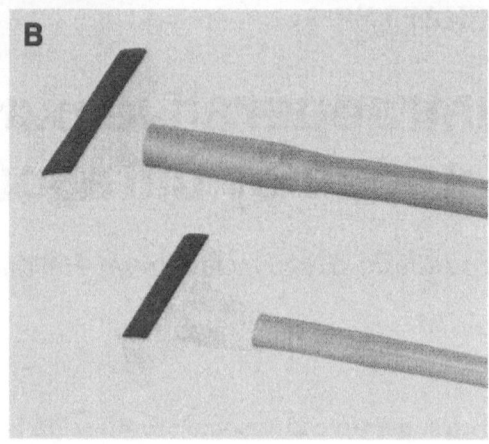

Fig. 9-1. Intraoperative ultrasonic probes for the liver
A. The present probe with puncture adaptor.
B. The upper probe is the same probe as in Fig. A and the lower is a 7.5 MHz probe 5.7 cm long and an ultrasonically visualized field of 3.8 cm in depth.

Biopsy of mass lesions in the liver

With intraoperative ultrasonic examination, many small nodules not detected by the preoperative examinations have been found. The histological nature of small nodules less than 2 cm in diameter cannot be differentiated with ultrasound. Small regenerative or hyperplastic nodules and small daughter nodules or intrahepatic metastases of hepatocellular carcinoma are all demonstrated as a simple low echo area. Therefore, aspiration cytology or needle biopsy is indicated. As small nodules deep in the liver are invisible and nonpalpable, ultrasonic guidance is necessary. The techniques of aspiration cytology and needle biopsy are the same as with percutaneous puncture, but much simpler because the targets are close and patient's respiration can be stopped completely.

In patients with nodules less than 2 cm in diameter in the cirrhotic liver, thick needle biopsy is preferable to aspiration cytology. In such small hepatocellular carcinoma, Edmondson's grade I type of low grade malignancy is very often encountered. The percentage of Edmondson's grade I in small hepatomas is higher than in hepatomas of more than 2 cm diameter. Edmondson's grade I and II is found in 85% of patients with small hepatomas. Therefore, cytological diagnosis is very difficult in these patients.

After needle biopsy has been performed, before retracting the needle oxycell cotton or a small piece of gelatin sponge is packed in the sheath. Then the needle is removed leaving the oxycell cotton in the liver. When this procedure is followed, no bleeding will occur from the puncture site. Therefore, this maneuver can be recommended when an ordinary percutaneous biopsy is performed.

Puncture of the bile duct

In patients with postoperative bile duct stricture, identification of the bile duct from the inflammatory connective tissue is sometimes difficult. Ultrasound and ultrasonically guided puncture can easily locate the bile duct. In cases with non-resectable bile duct or gall bladder carcinoma, internal or external drainage is necessary. However, with carcinomatous invasion of the porta hepatis, it is very difficult to drain all parts of the intrahepatic bile duct system and it is impossible to know if there are non-drained intrahepatic bile ducts. With ultrasonically guided puncture technique, it is easy to puncture and drain all parts of the intrahepatic bile duct system. When performing the puncture through the diaphragmatic surface of the liver the procedure is the same as described in Chapter 8, but an 18 gauge teflon-sheathed needle, 5 cm long, is used for the puncture. Puncture through the hepatic hilum is also possible. In this situation, the operator's index finger is inserted in the porta hepatis and a suitable puncture site is identified by ultrasound. Then the bile duct is punctured through the hepatic hilum with a sheathed needle. After the inner needle has been removed, the rigid end of the spring guide wire is inserted and advanced through the liver parenchyma to the diaphragmatic surface. With this procedure, the drainage tube can be left in the right anterosuperior area duct from the diaphragmatic surface (Fig. 9-2).

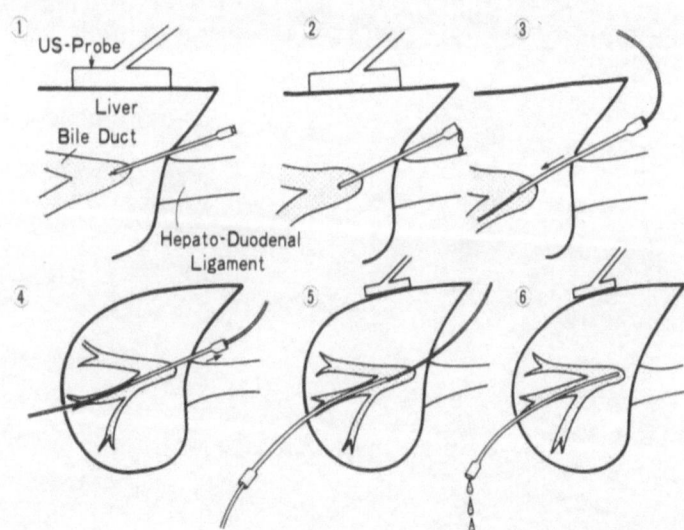

Fig. 9-2. External bile drainage through the hepatic hilum by intraoperative ultrasonically guided puncture technique
1) The right anterosuperior area duct is punctured from the hepatic hilum with an 18 gauge teflon sheathed needle 5 cm long.
2) Bile flows from the sheath.
3) The guide wire of 0.9 mm in diameter is inserted with the rigid end into the bile duct.
4) The guide wire penetrates the liver to the diaphragmatic surface and the sheath is pulled out.
5) The drainage tube is advanced over the guide wire to the obstructed portion of the bile duct.
6) After the guide wire is extracted, the position of the tip of the tube is estimated by ultrasound and the tube is fixed on the diaphragmatic surface of the liver.

With these techniques, it becomes easy to drain all parts of the intrahepatic bile duct system.

Hepatectomy and ultrasonically guided puncture

Ultrasonically guided partial hepatectomy: Makuuchi's subsegmentectomy

In patients with small hepatocellular carcinoma associated with liver cirrhosis, liver dysfunction will occur after surgery if hepatic lobectomy is performed. Therefore, partial hepatectomy is indicated. However, even a small hepatoma may invade the portal venous branches and tumor cells spread into its portal area. These tumor cells will probably give rise to daughter nodules. In our experience with small hepatomas, microscopic tumor thrombi were found in 73% of 62 resected specimens. Therefore, total resection of the portal area which contains the tumor is indicated (Makuuchi 1983).

In order to resect the portal area containing the tumor, accurate mapping of that area is required. For this purpose, the portal venous branch which contains the tumor is punctured under ultrasonic guidance and dye is injected (Fig. 9-3). If the tumor is located in the intersegmental or the intersubsegmental area and is fed by two portal venous branches, they are both punctured – first the

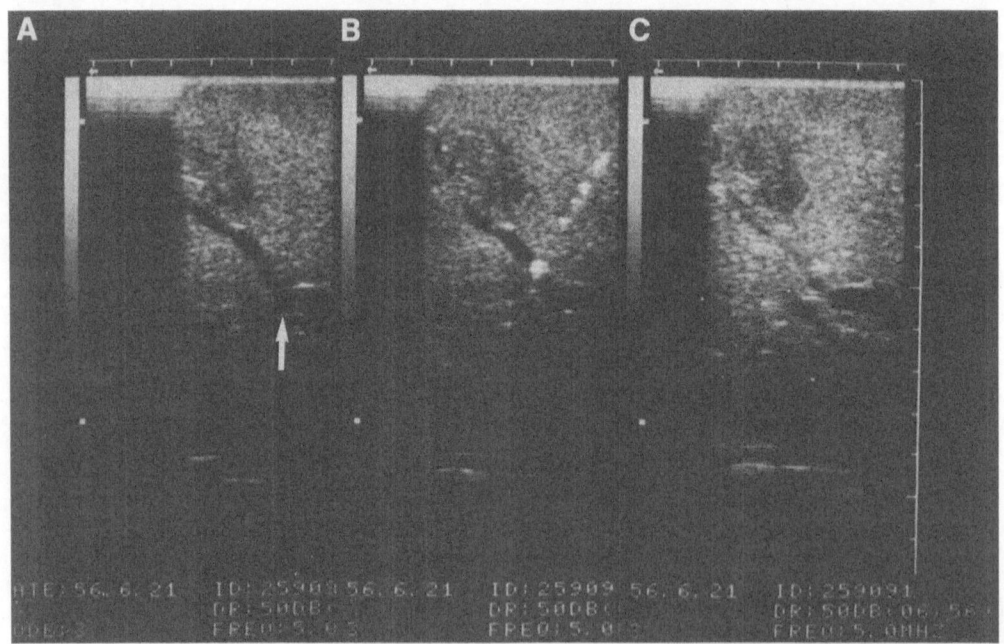

Fig. 9-3. Ultrasonically guided puncture of a portal venous branch and injection of dye
A. A small hepatocellular carcinoma and the feeding portal venous branch are demonstrated. The arrow indicates the confluence of the posteroinferior and the posterosuperior portal venous branches. B. The needle is inserted into the posterosuperior portal venous branch. The whole length of the needle is clearly seen.
C. The sonogram obtained during injection of dye. Due to microbubbles in the dye, the posterosuperior portal venous branch is obscured and the echo level of the liver parenchyma is increased.

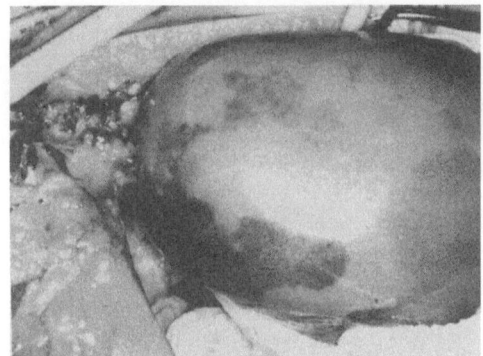

Fig. 9-4. Liver staining
The surface of the liver after the right anterosuperior portal venous branch is punctured and blue dye is injected. The anterosuperior area is totally stained. The rectangular defect of staining is due to compression of the liver surface by the probe.

dorsal and then the ventral vessel. With this technique, the liver surface stains blue (Fig. 9-4) and the stained area is marked with electrocautery.

The feeding portal venous branch of the tumor should be ligated at its confluence. During liver transection, a considerable amount of bleeding usually occurs and there is no land mark in the liver. Therefore, it is rather difficult to identify the ligation points of the vessels in the liver. In order to identify these points, blue dye is injected into the liver parenchyma just in front of the vessels to be ligated (Fig. 9-5). With this technique, it is easier to identify the ligating

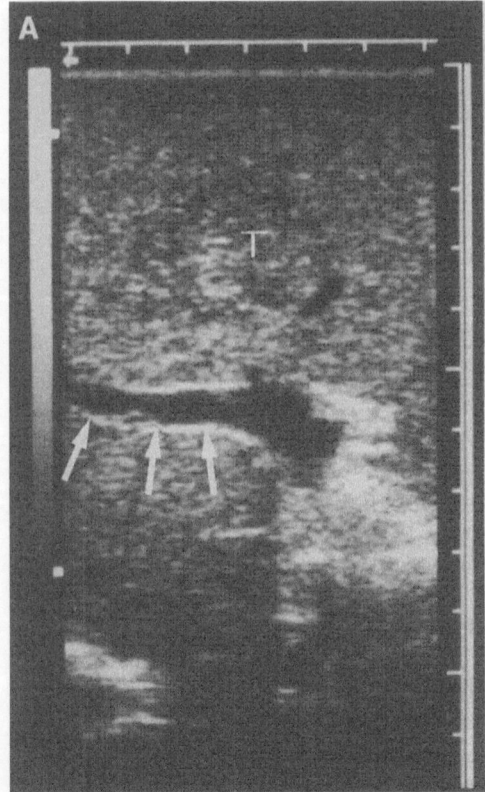
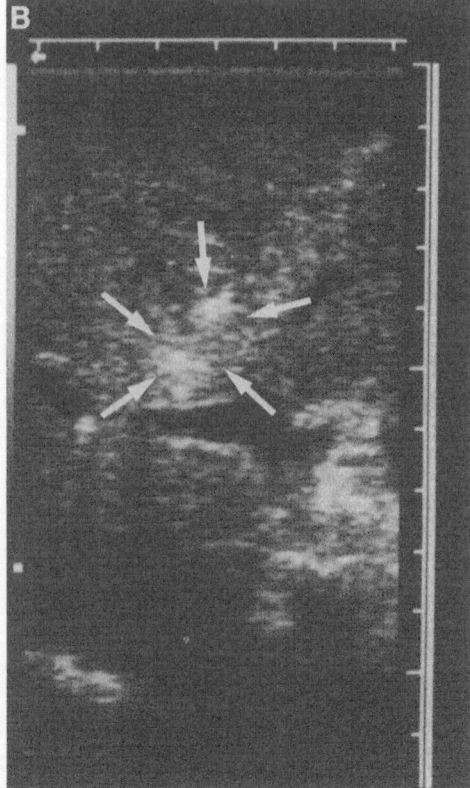

Fig. 9-5. Sonograms of tattoo of the liver
A. A tumor (T) and the posterosuperior portal venous branch (arrows) are demonstrated.
B. In front of the confluence of the posterosuperior venous branch with the main posterior branch, an echogenic mass (arrows) is seen. This mass is injected with dye for the tattoo.

point of the vessels because the dye marker is clearly seen during transection. We call this technique "Tattoo" of the liver. Fig. 9-6 shows the cut surface of the resected specimen and the tattoo is clearly seen near the portal venous branch.

These two puncture techniques are very important points in Makuuchi's subsegmentectomy. Further these two procedures are useful in standard hepatectomy or enucleation of deep tumors. Since 1979, we have performed 60 subsegmentectomies, in almost all cases in patients with cirrhosis. All patients recovered and were discharged from hospital.

Fig. 9-6. Resected specimen
The resected specimen of the anterosuperior area. Blue dye is seen on the cut surface and the portal venous branch (arrow) is just on the right of the tattoo.

References

Engel I A, Voorhies R M, Schneider M, Fraser R A R, Kazam E. Neurosurgical intraoperative ultrasound: tumor localization and characterization. In: Lerski R A, Morley P, eds. Ultrasound 82. Oxford: Pergamon Press, 1983: 353.

Sahn D J, Brandt W T, Barratt-Boyes B et al. A new technique for evaluation of coronary atherosclerosis by ultrasound scanning of the heart in open heart surgery. *Ultrasound Med Biol* 1982; 8(Suppl): 167.

Makuuchi M, Hasegawa H, Yamazaki S. Intraoperative ultrasonic examination for hepatectomy. *Jpn J Clin Oncol* 1981; 11: 367.

Lane R J, Glazer G. Intra-operative B-mode ultrasound scanning of the extra-hepatic biliary system and pancreas. *Lancet* 1980; 2: 334.

Sigel B, Coelho J C U, Spigos D G et al. Real-time ultrasonography during biliary surgery. *Radiology* 1980; 137: 531–3.

Cook J. H III, Lyton B. Intraoperative localization of renal calculi during nephrolithotomy by ultrasound scanning. *J Urol* 1977; 117: 543.

Lane R J. Intraoperative B-mode Scanning. *J Clin Ultrasound* 1980; 8: 427.

Makuuchi M, Kamiya K, Sugiura M, Wada T, Muroi T. Ultrasonic examination by electronic scanning during operation. *Proc Jpn Soc Ultrason Med* 1977; 32: 129 (Eng. abstract).

Makuuchi M, Hasegawa H, Yamazaki Y et al. Newly devised probe for intraoperative ultrasonic examination. Image Technology & Information Display. *Medical* 1979; 11 (18): 1167 (in Jpn).

Makuuchi M, Hasegawa H, Yamazaki S. Intraoperative ultrasonic examination for hepatectomy. Proceedings of the 26th annual convention of the American Institute of Ultrasound in Medicine. 1981: 319.

Makuuchi M, Hasegawa H, Yamazaki S et al. Differences in clinicopathological features and clinical diagnosis of small hepatocellular carcinoma among each diameters less than 5 cm and its treatment; systematic subsegmentectomy. *Acta Hepatol Jpn* 1983; 24: 1466 (in Jpn).

CHAPTER 10

Ultrasonically guided percutaneous nephropyelostomy

Per G. Lindgren & Anders Hemmingsson

Percutaneous nephropyelostomy (PNS) was rarely used for drainage of the renal pelvis and ureter until 1974 when a new technique was introduced, with the insertion of a polyethylene catheter on a trocar into the renal pelvis under fluoroscopic control (Almgård & Fernström).

For more accurate puncturing it is now well known that ultrasound is prefarable to fluoroscopy as it is a three-dimensional technique. The first ultrasonically guided nephrostomy was performed 1974 (Pedersen) and ultrasonic guidance has since become an established technique. The puncture of the renal pelvis is ideally done with the aid of a dynamic sector scanner, as the transducer requires only a small area of skin contact and is easily maneuvered (Lindgren).

It has proved possible to introduce a soft balloon catheter after 2 days of dilcatation. This has led to an increased use of PNS and in the following a technique will be described which has been used successfully in 358 patients.

Method

The patient is placed in the prone position on a fluoroscopic table and, with the aid of a sector scanner, the puncture site is located so that the patient can lie comfortably on his/her back when the catheter is in place.

The puncture site is usually below the 12th rib but occasionally a site between the lower ribs is used. When PNS is performed in a transplanted kidney the patient is placed in the supine position and the puncture site is chosen so that the peritoneum will not be penetrated.

The puncture is performed under sterile conditions. The ultrasound transducer is "sterilized" by draping it with a sterile glove and a needle guide unit is mounted on the scanner (Fig. 10–1). In the needle guide unit there are three different sized grooves for needles. The needle path is marked with guidelines on the monitor (Fig. 10–2). This makes it possible to *predetermine* the pathway of the needle exactly. During the puncture procedure the needle can be seen on the monitor.

A local anesthetic is injected into the tissues through a 20-gauge needle, from the skin surface to the site of penetration of the renal pelvis. With the tip of the needle in the renal pelvis, contrast medium is injected. With the use of the described ultrasonically guided puncture technique it is thus possible to inject a local anesthetic into the exact area

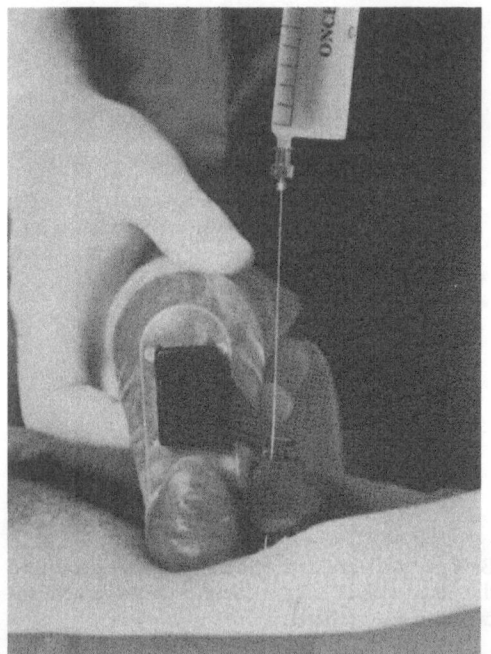

pressure and simultaneous rotation. The rotation makes it easier to penetrate the renal capsule (Fig. 10-3B). When the catheter has been placed with its tip in the renal pelvis or the upper part of the ureter, it is fixed to the skin with a suture and ensures good draining during the first 2 days. After 2 days, the steel-braided catheter can easily be changed to a soft balloon catheter (Foley No. 12) threaded over a P 205 guide wire. As the balloon permits internal fixation of the catheter, no skin suture is needed (Fig. 10-4). If long-term external drainage is planned, the balloon catheter is exchanged every 2nd month to avoid deposition of stone material in the catheter.

Fig. 10-1. Puncture transducer
The transducer is covered with a specially made sterile "glove". The needle guide unit is mounted on the transducer and the needle is introduced through one of the grooves.

where the catheter is to be introduced. A small incision is then made in the skin and, again using the ultrasound technique, a 16-gauge needle is inserted into the renal pelvis.

To increase the visibility of the needle, a special mandrin can be used, or the guide wire can be kept in place inside the needle (Fig. 10-2).

When the guide wire (P 205, D 1.19 mm) has been introduced, the needle is withdrawn and replaced by a polyethylene catheter (ID 1.8, OD 2.8 mm) under fluoroscopic control. Over this, a steel-braided polyethylene catheter (ID 2.9, OD 4.6 mm) with a radiopaque nylon tip is threaded (Fig. 10–3A). The catheter is introduced under moderate

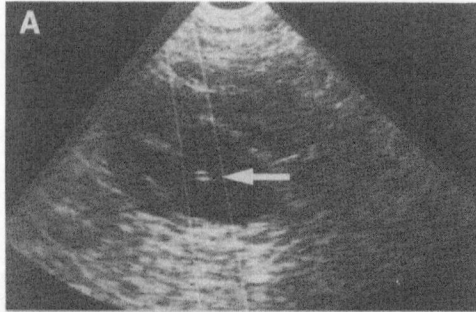

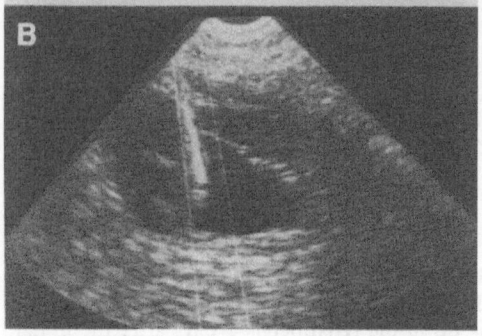

Fig. 10-2. Needle tip visualization
A: The needle path is represented by the area between the electronic lines. Between the two lines the tip of the needle is seen (arrow) placed in the dilated renal pelvis.
B: The needle is better visualized with a guide wire inside.

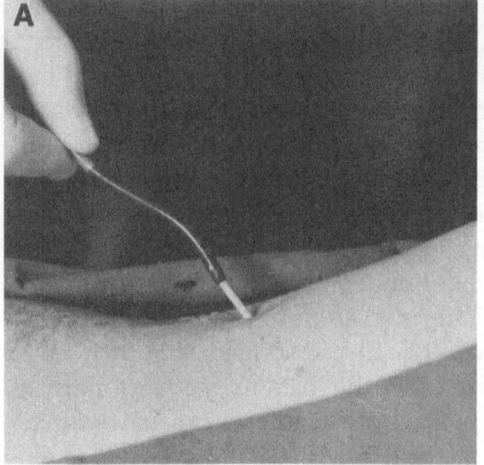

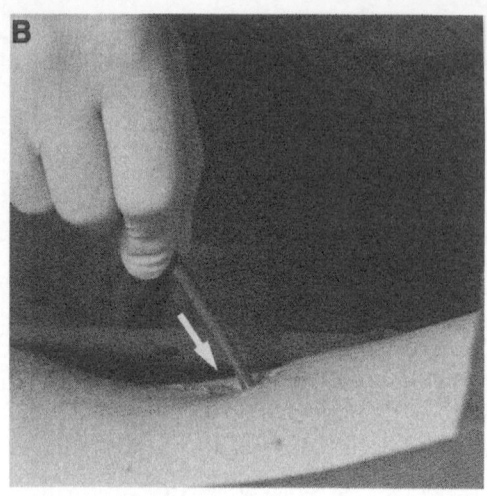

Fig. 10-3. Insertion of special catheter
A: A steel-braided polyethylene catheter (OD 4.6 mm) is threaded over the guide wire and the polyethylene catheter.

B: The steel-braided catheter is introduced into the renal pelvis with moderate pressure and rotation of the catheter.

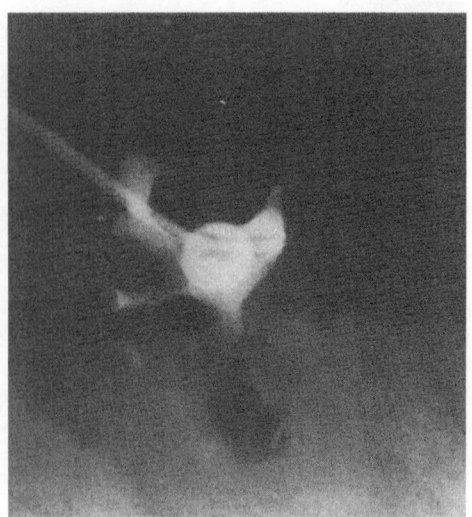

Fig. 10-4. Balloon catheter in place
The balloon catheter (Foley No. 12) is placed with the tip in the renal pelvis 2 days after the initial puncture.

place. Especially when prolonged or permanent drainage is needed, internal fixation made possible with the Foley catheter is an advantage. In our experience this technique is to be preferred to one-step procedures with dilation of the track with coaxial dilators or balloon catheters (LeRoy et al.).

Material

In our material of 358 patients the major causes of obstruction of the ureter were pelvic neoplasms of different types. The most common were tumors of the bladder (37%) and prostate (17%) and gynecological tumors (13%). Other reasons for PNS were injuries to the ureter during surgery, pregnancy, rectal carcinoma with overgrowth of the tumor, malignant lymphomas and chronic cystitis.

Other types of catheter can be used such as the pig-tail catheter. In our opinion this catheter, as well as other types, needs a skin-suture to be kept in

In 144 patients bilateral PNS was performed and in all of them it was carried out on both sides on the same occasion. The time required for bilateral PNS was usually less than 45 minutes. In 2 cases PNS was carried out on the same occasion on the dilated homolateral upper and lower renal pelvis with separate ureters.

In 12 cases with surgical lesions of the ureter the indication for PNS was leakage of urine. In all these cases at least a small amount of fluid in the renal pelvis could be demonstrated ultrasonically and the pelvis therefore punctured.

Pyonephrosis was found in 9 cases as a complication to obstruction of the distal part of the ureter. In all these cases the renal pelvis could be drained with the coarse steel-braided catheter which was left in place for 4–8 days to secure the drainage before it was changed to a soft balloon catheter.

Two small children were treated with PNS. They had hydronephrosis and were clinically suspected of pyonephrosis. In both cases the urine was clear and the thin polyethylene catheter was left in the renal pelvis for 1–3 days prior to surgery.

In 7 patients PNS was done on transplanted kidneys with obstruction or leakage of urine at the site of the anastomosis.

Complications

No major complications occurred in our series.

The mortality rate of percutaneous nephrostomy has been reported to be 0.2% and the rate of significant complications, all of them hemorrhages, 4% (Stables). Minor complications such as dislodgement of the catheter and debris obstruction of the catheter by blood clots or debris are reported in about 15% of cases. Transient hematuria is not regarded as a complication. Small perinephric hematomas have been found in 8% (Stables et al.).

In our series displacement of the steel-braided catheter occurred in 5 patients. Dislodgement of the Foley catheter occurred in 28 cases. In the majority of these cases there was a small leakage of fluid from the balloon so that it decreased in size, finally permitting the catheter to slide out. In many of these cases a guide wire could be placed in the renal pelvis simply by using the preformed track. In these cases it is often easier to replace the guide wire if contrast medium is injected into the canal to visualize the whole canal and the renal pelvis. When the guide wire is in place, the soft balloon catheter can be threaded over it. This procedure is often possible if no more than 12–14 hours have elapsed since dislodgement of the catheter. In other cases a new ultrasonically guided PNS must be undertaken. One of our cases has had permanent bilateral catheters for 6 years.

In 23 patients obstruction of the Foley catheter occurred because of deposits in it. In most of these cases a guide wire could be introduced into the Foley catheter which was then exchanged for a new one.

References

Almgård L E, Fernström I. Percutaneous nephropyelostomy. *Acta Radiol Diagnosis* 1974; 15: 288.

Pedersen J F. Percutaneous nephrostomy guided by ultrasound. *J Urol* 1974; 112: 157.

Pedersen J F, Cowan D F, Kvist Kristensen J, Holm H H, Hancke S, Jensen F. Ultrasonically-guided percutaneous nephrostomy. Report of 24 cases. *Radiology* 1976; 119: 429.

Stables D P. Percutaneous nephrostomy: techniques, indications and results. *Urol Clin North Am* 1982; 9: 15.

Lindgren P G, Hemmingsson A. Percutaneous nephropyelostomy. *Acta Radiol Diagnosis* 1980; 21: 759.

Lindgren P G. Ultrasonically guided punctures. A modified technic. *Radiology* 1980; 137: 235.

Heckemann R, Seidel K J. The sonographic appearance and contrast enhancement of puncture needles. *J Clin Ultrasound* 1983; 11: 165.

Elyaderani M K, Dorn J S, Gabriele O F. Percutaneous nephrostomy utilizing a pigtall catheter: A new technique. *Radiology* 1979; 132: 750.

LeRoy A J, May G R, Segura J W, Patterson D E, McGough P F. Rapid dilatation of percutaneous nephrostomy tracks. *Am J Roentgenol* 1984; 142: 355.

Stables D P, Ginsberg N J, Johnson M L. Percutaneous nephrostomy: A series and review of the literature. *Am J Roentgenol* 1978; 130: 75.

Percutaneous nephrolithotomy

P. Alken

Percutaneous nephrolithotomy (PNL) as it is performed today is a combination of several techniques described some years ago. A synthesis of these techniques including mechanical stone removal, chemolysis, and ultrasound disintegration was first presented in 1980 (Alken & Altwein). These initial experiences suggested that PNL could become an alternative to open stone surgery. Since then, more than 2000 such procedures have been performed worldwide. Several modifications have been described leading to a comparable success rate of up to 95%.

Percutaneous renal anatomy and access

The optimal access to the collecting system is established through a puncture that lies in extension of the long axis of a calyx. Only the posterior row of calyces offers such a safe entry into the collecting system. Most frequently an access through the posterior middle or lower calyx is chosen. The posterior part of the lower renal pole, which is relatively devoid of larger arteries, is the safest area of entry. *The peripheral puncture* is a transpapillary puncture. Within this area of the renal parenchyma there

are no larger arteries. The transparenchymal tract leads into the collecting system where it is affixed to the renal papilla. Extravasations are rarely observed during the subsequent intrarenal instrumentation. As the calyx forms part of the tract, additional space is gained within the collecting system for inspection and instrumentation (Fig. 11-1). The risk of losing the tract is slight. *The central puncture* takes a transparenchymal route next to the papilla through the renal sinus into the collecting system. The risk of lesions to intralobar and segmental arteries is high. Extravasations are frequent, because the pelvic wall does not seal off the tract. There is only a small space for a safe intrarenal instrumentation and the risk of losing the tract is high. A central puncture should therefore be avoided.

Patient selection and preparation

Patients with purely controlled hypertension or coagulation disorders should be excluded from PNL. The patient should be cooperative, especially when several sessions are required. In cases with acute obstruction and urinary tract infection preliminary decompression by

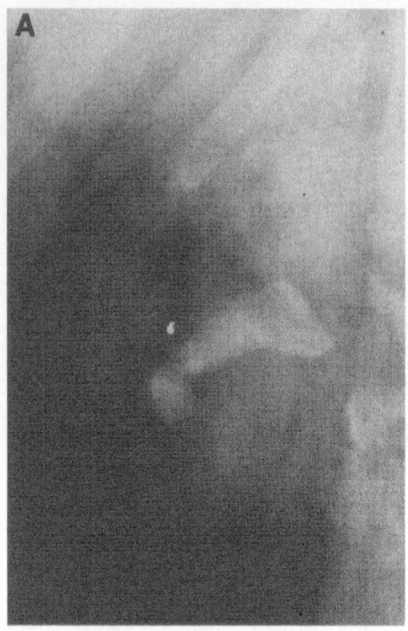

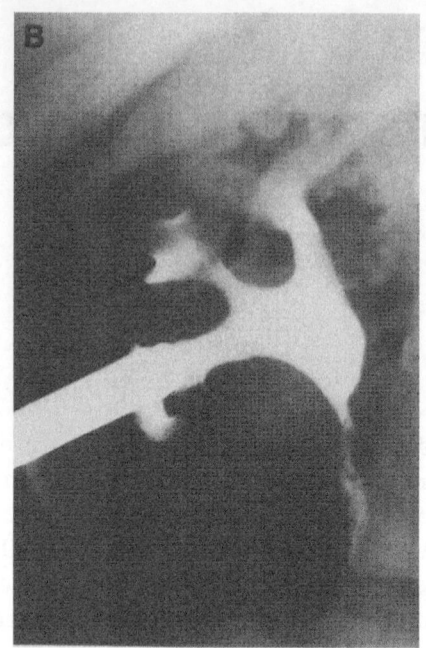

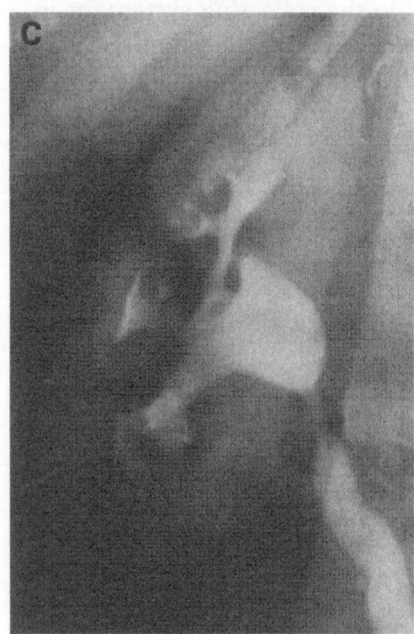

Fig. 11-1. Percutaneous Nephrolithotomy
A: Pyelocalyceal stone, plain film. B: IVP. C: Complete stone removal by ultrasound disintegration in a 1-step procedure. Procedural time 70 min. Final antegrade pyelography.

percutaneous nephrostomy and antibiotic treatment should precede intrarenal instrumentation. Selection of the patients depends on the localization of the stone, its accessibility via a straight channel, the ease with which the puncture can be performed, the availability of appropriate instruments, and experience. Even though in our own series it has never been necessary to perform an emergency operation immediately after an unsuccessful percutaneous trial, patients are routinely prepared as for open surgery. This is not necessary if a multiple-step procedure is performed exclusively under local anesthesia.

Anesthesia

The form of anesthesia largely depends on the anesthesiologist and whether a 1- or a multiple-step procedure is planned. It is possible to perform all procedures under local anesthesia, but some patients will experience pain which will render the instrumentation difficult or unsuccessful. The most frequently applied forms of anesthesia are: general anesthesia, intravenous sedation in combination with local anesthesia and regional anesthesia. Peridural anesthesia was effectively performed in most of our patients.

One-step or x-step procedure

The choice is open, as large series show comparable success rates independent of the number of sessions. The experience in our first series has shown that it takes 4–5 days until a mature channel has formed which allows the safe introduction of instruments with a minimal risk of bleeding and optimal conditions for intrarenal instrumentation. But the nephrostomy tube may dislodge within that time and for the patient the multiple-step procedure is less attractive than a 1-step procedure. The so-called immediate 2-step procedure (Clayman & Casteneda-Zuniga 1984) with puncture on day 1 and dilation and stone removal on day 2 only offers logistic advantages as a sometimes lengthy procedure is divided into 2 steps. Again, tubes may dislodge and the tract is as unstable as in a 1-step procedure.

In our department, a 1-step procedure is performed whenever possible. The whole procedure is exclusively done by the urologist, and special instruments have been designed to minimize the problems of working through an acutely established tract. The 1-step procedure is described in the following sections.

Puncture

All percutaneous procedures are performed under sterile conditions. The patient lies prone on an X-ray table and is swathed in plastic self-adhesive drapes. Especially when the collecting system is not dilated, a drip infusion IVP facilitates the puncture. Under fluoroscopic control the direction of the long axis of the target calyx is marked on the patient's back. Thereby the plane of puncture is defined. By subsequent ultrasound scanning within that plane the optimal puncture site, lying in extension on the long axis of the calyx is selected. The puncture is performed under combined ultrasound and fluoroscopic control. Once the needle ((6761) 1.6 mm Ø Becton Dickinson Rutherford NJ 07070 USA) has reached the calyx, contrast dye is injected and all further steps are performed under fluoroscopic control.

Dilation

If chemolysis is planned dilation up to 16 french is usually sufficient to introduce an outflow- and an inflow-catheter in a parallel fashion. Stone removal by extraction or disintegration requires a channel up to 26 french.

A J-guide wire is introduced through the teflon sleeve of the needle and the first dilation steps up to 7, 9 or 11

French are performed with a Hoffmann-Desilet Catheter Introducer Set ((1167, 07-11) Vygon BP 795440 Ecouen France) consisting of bougie and sheath. Upon removal of the dilator a second safety guide-wire can be introduced through the plastic sheath. All further dilation steps are performed with a metallic telescope dilation set (Alken 1981), consisting of a 60 cm long guide rod and 6 metallic dilators of 9, 12, 15, 18, 21 and 24 french diameter (Karl Storz KG D-7200 Tuttlingen, FRG (27090 A)). Through the plastic sheath, the guide rod is introduced into the collecting system under X-ray control. If the hollow version of the guide rod is used it is introduced over a J-guide wire. In the latter case an 11 french plastic sheath must be in place. The plastic sheath is then removed and 1 metallic dilator is advanced over the other. A prerequisite for the use of the telescope dilators is a straight course tract. The advantage of the telescope dilators is that the original straight puncture path is maintained during the dilation and that the channel is sealed off during the whole dilation procedure with minimal bleeding into the collecting system and to the outside.

Intrarenal instrumentation

Sheats of the 18, 21 or 26 French nephroscopes are introduced over the telescope dilators like an additional dilator and upon withdrawal of the inlying dilators, the nephroscope is introduced.Diuresis is induced by either the contrast dye given at the beginning of the procedure or by a mannitol infusion. The kidney is less susceptible to pyelorenal reflux under these conditions. Immediately prior to the endoscopic inspection of the collecting system, contrast dye is injected to rule out any major extravasation which in turn would be a reason for postponing intrarenal instrumentation. Physiological NaCl- solution is used for irrigation of the collecting system.

Stone removal

Chemolysis is only applicable in a small percentage of stones composed of either uric acid, cystine or struvite. The solutions used in these cases are $NaHCO_3$ 1.1%, pH 8.7 for uric acid stones, N-acetyl-cystein 2.5%, combined with $NaHCO_3$ 1.1% for cystine stones, and Renacidine®, 10%, pH 3.9 for struvite stones. The patients should be under antibiotic coverage during the procedure, the height of the fluid reservoir should be adjusted to less than 30 cm above kidney level and any outflow obstruction must be ruled out. Chemolysis is a time-consuming procedure and therefore it is currently only applied as an additional maneuver for residual stones and only after partially successful surgery, percutaneous nephrolithotomy or shock wave therapy. *Extraction* of small stones may be accomplished using various baskets, forceps or graspers through the nephroscope sheath. Extraction of larger stones through the bare channel may lead to bleeding that is especially troublesome if the stone is lost within the tract. Therefore stones that do not fit through the nephroscope sheath are preferably disintegrated by mechanical devices, ultrasound or electrohydraulic disintegration. The problem of *disintegration procedures* is the generation of fragments that may pass into the ureter or remote calyces, where they may be difficult or impossible to reach. This risk of iatrogenic residual stones can only be kept low by a controlled disintegration or fragmentation of the stones.

Controlled disintegration is performed with an ultrasound probe. Through the hollow probe, which is connected to a pump, small particles are immediately suctioned out during the disintegration process. If fragments are produced they should be immediately disintegrated further or removed by exchanging the ultrasound probe for forceps. Thus, ideally, the stone is gradually reduced in size and no fragments are left in place during the disintegration process. The technique is most efficient if the stone is brittle. *Controlled fragmentation* may be performed by electrohydraulic disintegration. This technique may be applied when the stone is very hard and has a tendency to break up into fragments instead of many small pieces.

The fragments then have to be removed with forceps. Ultrasound and electrohydraulic disintegration have been used by several authors with comparable success rates. If the ultrasound probe is inadvertently brought into contact with the pelvic wall there will be no sequelae. In the same situation the electrohydraulic probe will lead to a perforation of the pelvic wall. A safety distance of 5 mm from the tip of the electrohydraulic probe to the lens system of the nephroscope is necessary to avoid damage of the lens.

Postprocedural care

The procedure is terminated with an inspection of all accessible calyces with a 0, 30 or 70 degree telescope and plain film. Residual stones may be flushed out with preshaped angiocatheters or removed with the help of a flexible nephroscope. In the case of no residual stones being documented, a nephrostomy tube is placed to secure postprocedural urinary drainage and hemostasis. The size of this nephrostomy tube is selected according to the degree of bleeding. In the 1-step procedure we introduce a nephrostomy tube of the same diameter as the instrument that has been used for intrarenal instrumentation. This guarantees that bleeding is stopped by compression. Only if the bleeding tendency is slight is a small calibre catheter placed. Two techniques and 2 types of catheter may be used.

Introduction of the nephrostomy tube with a size equivalent to the nephroscope sheath. The inner diameter of the nephrostomy tube is probed with the telescope dilators. All dilators that fit into the nephrostomy tube are loaded on the guide rod and introduced into the collecting system through the nephroscope sheath. The nephroscope sheath is removed, the tube is introduced over the dilators and by pulling on the guide rod the inlying dilators are removed. *Introduction of a Foley catheter of 22 French diameter.* All telescope dilators are reintroduced through the nephroscope sheath. The sheath is exchanged against a slotted canula of 26 french outer diameter. By pulling on the guide rod the inlying dilators are removed and a 22 french Foley catheter is placed into the collecting system through the slotted canula which is thereafter removed.

The advantage of a Foley catheter is that the balloon keeps the catheter in place and tube dislodgement that may be seen in approximately 15% of cases

81

when a simple nephrostomy tube is used can be prevented. But depending on the configuration of the renal pelvis it may not always be possible to inflate the balloon to a proper size without blocking the calyces with consequent obstruction and complications. Any type of catheter should be fixed by a skin suture. The tubes are left in place for 4 days or may be removed earlier if there is little bleeding tendency.

Secondary procedures

Secondary procedures may become necessary because of limited view due to bleeding or because of residual stones. It is our policy to wait 4 to 5 days until a mature channel has formed.

Results

Percutaneous stone removal was attempted in 266 renal units. An operatively established nephrostomy served as access in 21 cases. Failure to establish a proper percutaneous access or residual stones of a size too large for spontaneous passage necessitated surgery in 16 patients. An additional 2 patients were treated with shock waves for the residual stones. *Chemolysis* alone or in combination with endourological procedures was successful in 18 of 21 cases. But an average of 22 days and a maximum of 60 days of treatment were necessary for complete stone dissolution. No complications were met in the series. *Ultrasound disintegration* was the most frequently applied technique of stone removal, the electrohydraulic lithotrite being used in only 7 cases. The 1-step procedure resulted in immediate success in 71% of 165 cases. In 1 patient, 4 sessions were necessary for complete stone removal. The average time for the 1-step procedure was 75 minutes for a simple stone extraction and 111 minutes if the stone required intrarenal disintegration. Average time of hospitalization was 7–9 days in those patients with at 1- or 2-step procedure and 12 days if more than 2 sessions were necessary.

The residual stone rate

The residual stone rate was 9 of 21 in those patients with an operatively established nephrostomy and 16 out of 232 successful procedures in patients with a primary percutaneous access.

Complications

Ten early and 4 late complications were met. Acute pyelonephritis, bleeding, or extravasation were usually managed conservatively. Of the 3 major complications – perforation of the colon, UP-junction obstruction and arterial bleeding – the colon perforation was managed conservatively by turning the percutaneous nephrostomy into a percutaneous colostomy with the tube later being removed without any sequelae. The arterial bleeding required a transarterial superselective embolization and the patient with the UP-junction obstruction required surgical intervention.

Discussion

PNL has completely replaced open stone surgery in cases of easily accessible caliceal or pelvic stones. In our own department with a large turnover of staghorn stone surgery approximately 50% of all stones were treated by a percutaneous access. Since the acquisition of the extracorporeal shock wave lithotriptor in our department in January 1984, a change in indications has taken place. The easy stones that were formerly managed by PNL are now treated with the shock wave machine but the frequency of percutaneous procedures has not diminished since January 1984. Patients with radiolucent stones that do not allow a proper focusing in the shock wave machine, and patients with equivocal outflow obstruction, due to narrow calyceal necks or a narrow UP-junction, are still exclusively treated by percutaneous procedures. If the stone mass is very large, e.g. in staghorn stones, an exclusive treatment by shock wave therapy frequently leads to an obstruction and subsequent infection because of a hampered passage of the multiple fragments through the ureter. In these cases particularly, a combined procedure by a primary percutaneous approach and a secondary shock wave treatment of the residual stones has proved to be very effective. Thus, even with the advent of the shock wave lithotriptor, PNL still has a place in the treatment of renal stones, especially for complicated cases.

References

Alken P, Altwein J E. Die Perkutane Nephrolitholapaxie. *Verh Dtsch Ges Urol*, 31. Tagung. Berlin, Heidelberg, New York: Springer Verlag, 1980: 109.

Alken P, Hutschenreiter G, Günther R, Marberger M. Percutaneous stone manipulation. *J Urol* 1981; 125: 463.

Alken P, Günther R, Thüroff J. Percutaneous Nephrolithotomy – A Routine procedure? *Br J Urol* 1983; (Suppl.) 1.

Clayman R V, Casteneda-Zuniga W. Nephrolithotomy: Percutaneous removal of renal calculi. *Urol Radiol* 1984; 6: 95.

Alken P. Teleskopbougierset zur perkutanen Nephrostomie. *Aktuelle Urologie* 1981; 12: 216.

Alken P, Hutschenreiter G, Günther R. Percutaneous kidney stone removal. *Eur Urol* 1982; 8: 304.

Weirich W, Frohneberg D, Ackermann D, Alken P. Praktische Erfahrungen mit der antegraden lokalen Chemolyse von Struvit/Apatit-, Harnsäure- und Zystinsteinen in der Niere. *Urologe A* 1984; 23: 95.

Alken P. Percutaneous ultrasonic destruction of renal calculi. *Urol Clin North Am* 1982; 9: 145.

Alken P. Percutane Nephrolithomie. *Urologe A,* 1984; 23: 20.

Wickham J E A, Miller R A. Percutaneous Renal Surgery, Churchill Livingstone, Edinburgh, London, Melbourne, New York, 1983.

Clayman R V, Casteneda-Zuniga W. Techniques in Endourology, 1984.

Puncture of renal mass lesions

Niels Juul, Søren Torp-Pedersen & Svend Larsen

The ultrasonic accuracy in differentiating simple renal cysts from other masses is high and ultrasound has therefore become one of the primary diagnostic procedures in cases of renal mass lesions.

Solid renal masses

A solid renal mass may present with a break in the renal contour and often displacement of calyceal echoes. These masses differ in echogenicity from echo-rich to echo-poor (Fig. 12-1). However, a hypernephroma may be situated anywhere in the kidney and it is sometimes impossible from the ultrasound image to differentiate between such a tumor and a tumor of the renal pelvis (Figs. 12-2, 12-3). Neither is it possible from the ultrasound image to differentiate with certainty between malignant and benign tumors, as demonstrated by the appearance of the benign angiomyolipoma in Fig. 12-4.

In cases of severe chronic pyelonephritis or infarction of the kidney the irregular appearance of the organ may give rise to suspicion of a tumor (Figs. 12-5, 12-6). An intrarenal hematoma will appear cystic, but during or-

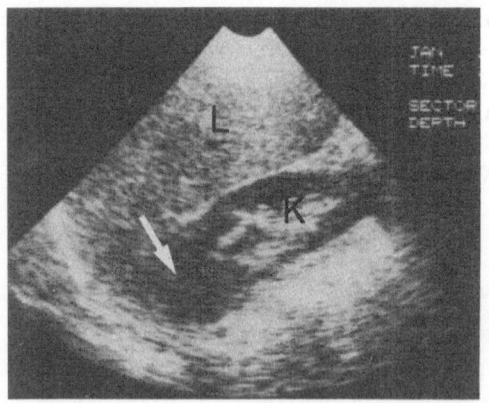

Fig. 12-1. Typical hypernephroma
Longitudinal scan of the right kidney showing a solid mass in the upper pole (arrow). L: Liver. K: Kidney. Operation revealed a hypernephroma.

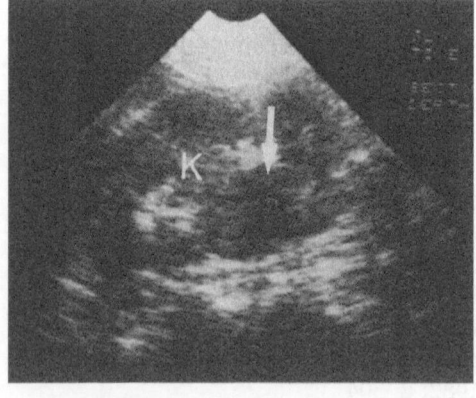

Fig. 12-2. Hypernephroma
Centrally placed solid tumor in the kidney (arrow). K: Kidney. Operation revealed a hypernephroma.

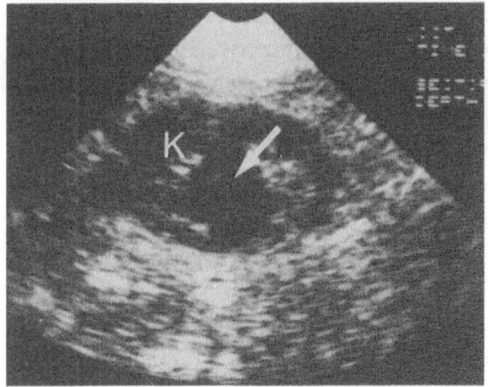

Fig. 12-3. Pelvic tumor
Centrally placed echo-poor tumor in the kidney (arrow). K: Kidney. Operation revealed a pelvic tumor.

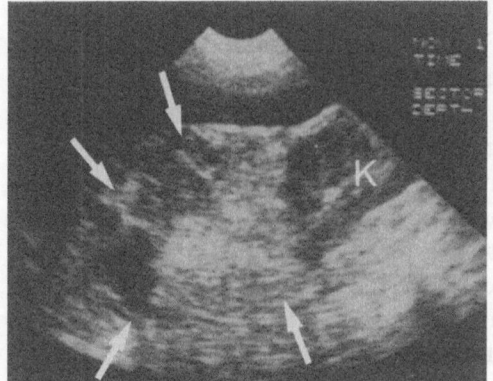

Fig. 12-4. Angiomyolipoma
Echo-rich irregular tumor of the upper kidney pole (arrow). K: Kidney. Operation revealed a benign angiomyolipoma.

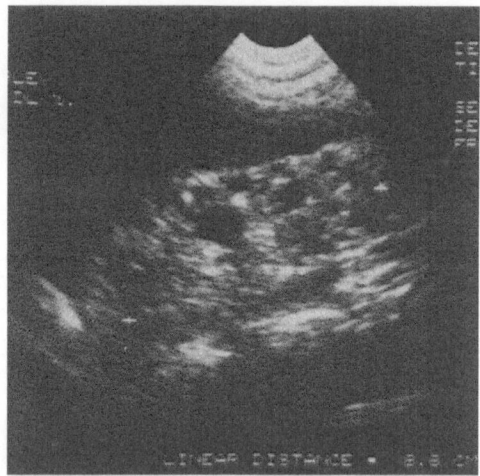

Fig. 12-5. Contracted kidney
Longitudinal scan of a contracted kidney (length 8.8 cm) due to chronic pyelonephritis.

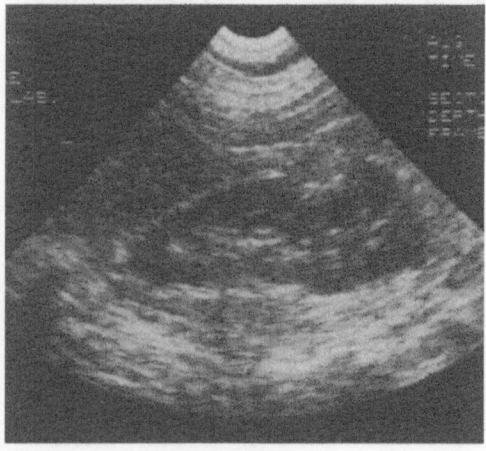

Fig. 12-6. Infarcted kidney
Longitudinal scan of a kidney with an irregular lower pole due to pyelonephritis and infarction.

ganization it will change its echogenicity and appear solid (Fig. 12-7). In these cases the clinical story will often, but not always, be decisive.

A displaced renal lobe (renal columnar hypertrophy) is typically a protrusion of parenchymal tissue in between the calyceal echoes. It is contiguous with the renal cortex, of the same echointensity and not larger than 3 cm

(Fig. 12-8). This condition too may be mistaken for a renal tumor.

Fine needle biopsy of solid renal masses

When a solid renal mass is disclosed or suspected an ultrasonically guided fine

85

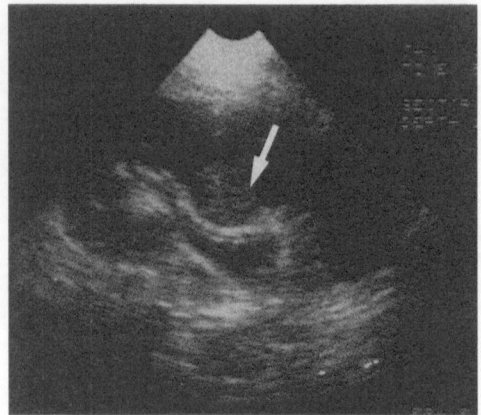

Fig. 12-7. Renal trauma
Centrally placed echo-rich renal mass due to organized hematoma following blunt renal trauma. Aspiration cytology and angiography without signs of malignancy.

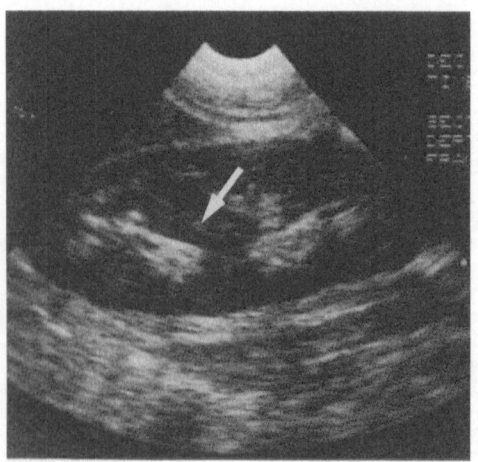

Fig. 12-8. Displaced renal lobe
Typical appearance of a displaced renal lobe (arrow).

needle biopsy should be considered. This procedure was first described in 1972 by Kristensen et al. The patient should be placed in a prone or at least decubitus position and the puncture should be performed from the back. Upper pole masses should be punctured in a cephalad direction to avoid traversing the pleura. The puncture and aspir-

ation techniques are described in Chapters 2 and 4. The procedure is safe and no patient preparations necessary, making it suitable for out-patients.

A series of 301 consecutive ultrasonically guided fine needle aspiration biopsies of solid renal masses was performed during a 12-year period (Table 12-1). The retrieval rate was 95%. A correct cytological answer was obtained in 82%. There were 25 false negative aspirates. The predictive value of a non-malignant aspirate was 71%. There were 14 false positives, and the predictive value of a malignant aspirate was 93%. The malignant masses were renal adenocarcinomas and urothelial tumors. Furthermore, 2 teratocarcinomas and 1 case of malignant lymphomatous infiltration were found. The benign lesions were inflammatory conditions, benign tumors such as adenomas, angiomyolipomas, and fibromas, together with a few normal anatomic variations (displaced renal lobe, dromedary kidney, etc.). The 14 false positives are shown in Table 12-2. The predictive value of a non-malignant aspirate is at the same level as the results obtained by puncture of other organs. However, the predictive value of a malignant aspirate is significantly lower than for any other abdominal organ, where false positives are extremely rare. The false positive aspirates were due to misinterpretations of the cytological material for a variety of reasons. Many different epithelial cells may be aspirated from a non-malignant lesion or even from a normal kidney. Histiocytes of inflammatory processes may cause diagnostic difficulties. Cells indistinguishable from well differentiated adenocarcinomas of clear cell type may be obtained from small

Table 12-1. Comparison between cytologic and final diagnoses

Final diagnosis		Cytologic diagnosis		
		Malignant	Benign	Insufficient
Malignant	218	185	25	8
Benign	83	14	61	8
Total	301	199	86	16

Predictive value of malignant aspirate: 93% (87–95).
Predictive value of non-malignant aspirate: 71% (60–80).
95% confidence limits in brackets.

Table 12-2. False positive aspirates

Inflammatory condition	6
Small, benign cyst	3
Adenoma	3
Cystic teratoma	1
Mesenchymoma	1
Total	14

benign cortical adenomas. Finally, cells aspirated from adrenal glands may be mistaken for tumor cells. All the false positive cytology answers were misinterpreted as rather well differentiated adenocarcinomas. Such a cytological answer should, in our opinion, be viewed with caution. A cytological diagnosis of, for instance, poorly differentiated clear cell carcinoma (Fig. 12-9) or urothelial carcinoma (Fig. 12-10) is much more reliable. Furthermore, a correct cytodiagnosis has been obtained preoperatively by Glenthøj & Partoft in 2 cases of angiomyolipomas. Radical nephrectomy could thus be avoided and substituted by the more conservative heminephrectomy. However, it seems questionable whether a fine needle aspiration biopsy is indicated in all patients with solid renal masses. Such a biopsy, on the other hand, may add important diagnostic information in equivocal cases, when angiography reveals an avascular mass, or is contraindicated when a small pelvic tumor or, e.g., an angiomyolipoma is suspected.

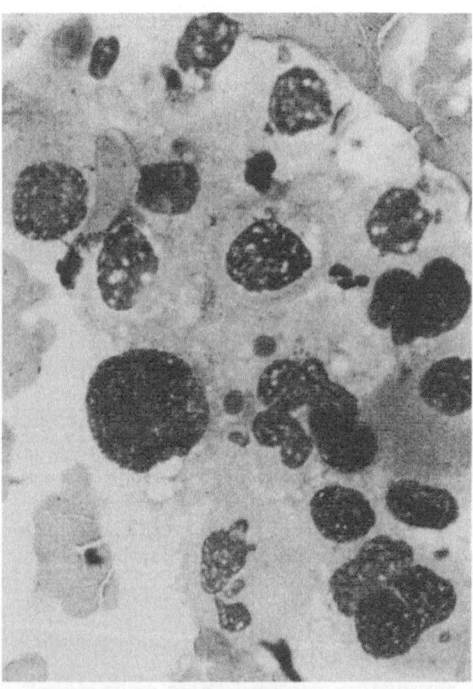

Fig. 12-9. Hypernephroma
Cytology showing malignant cells from a poorly differentiated hypernephroma. The malignant cells show a great variation in size, the contours are irregular and the chromatin shows a coarse meshwork. H & E, 500 ×.

However, the risk of a false positive cytological diagnosis should always be kept in mind.

Cystic renal masses

Cystic lesions in the kidneys are common and easily diagnosed ultrasonically down to a size of 1 cm. Most renal cysts

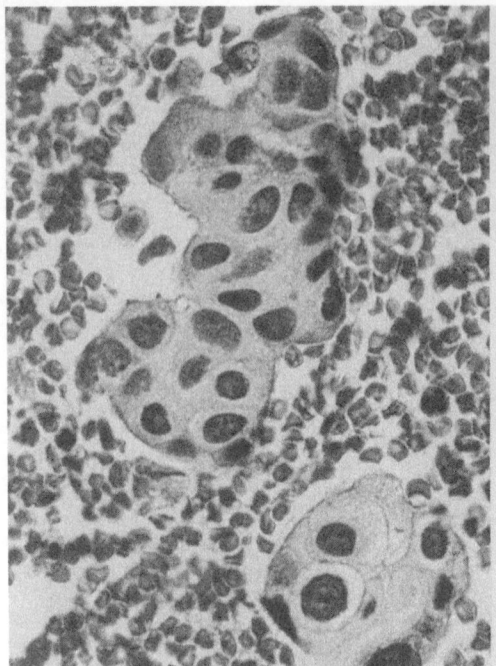

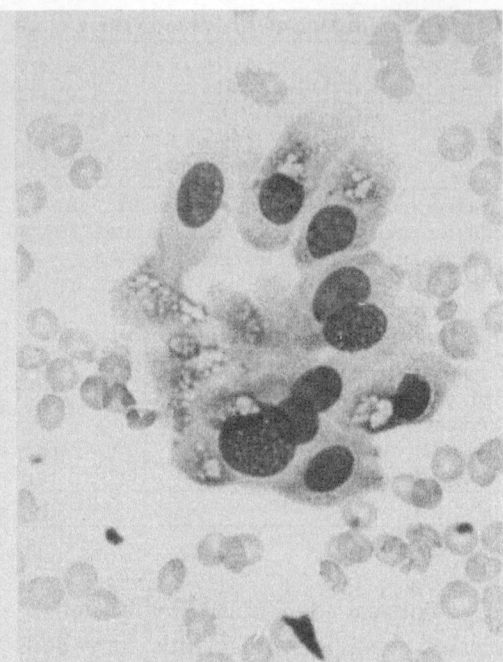

Fig. 12-10. Urothelial tumor
Cytology showing a sheet of cells from an urothelial tumor. Clumps of tumor cells are seen. The nuclei vary in size and the largest is about 2–3 times the normal size of the nucleus. The nucleus contour is irregular and the cytoplasm is reduced, making the image cell-rich. H & E, 500x.

Fig. 12-11. Renal tubular cells
Cytology showing a population of tubular cells from a normal kidney. As seen here the cells may vary in size, but the contour of the nuclei is regular and the chromatin shows a fine meshwork. H & E, 500 × .

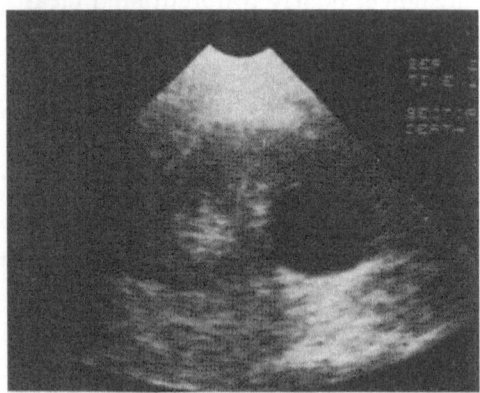

Fig. 12-12. Renal cyst
Simple cyst of the lower kidney pole. Note the round regular shape of the cyst and enhancement phenomenon.

are asymptomatic and their only significance is the differential diagnosis to renal carcinomas. Although malignancy of simple cystic lesions in the kidney is extremely rare, some authors advocate that all cysts should be treated. At least cysts with even slight irregularity or multiloculated cysts should always be punctured to exclude a renal carcinoma (Figs. 12-12, 12-13). Furthermore, centrally placed cysts, especially in obese patients may be misinterpreted and should be punctured to exclude a renal pelvic tumor.

The puncture technique is similar to that of solid renal lesions. The patient should be placed in the prone or decubitus position and cysts in the upper pole

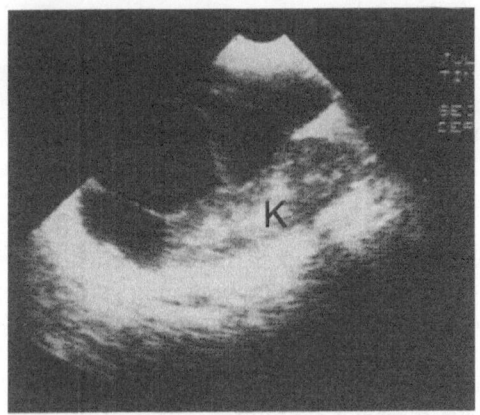

Fig. 12-13. Necrotic hypernephroma
Multiocular cystic renal mass with solid areas. K: Kidney. Operation revealed a hypernephroma with large necrotic areas.

should be punctured in a cephalad direction. Local anesthesia is applied and the cyst is punctured with a 1.2 mm (18 G) lumbar needle. The needle should be introduced in one rapid movement. When dealing with small, centrally placed cysts, puncture with a 0.6 mm (23 G) needle supported by an outer guide needle is advocated. The cyst should be emptied quantitatively.

Normally, renal cyst fluid is yellowish and clear. During aspiration it may turn red due to the puncture trauma. Occasionally the fluid is brownish, due to previous hemorrhage, and malignancy should be suspected. The aspirated material should be sent for cytological and chemical evaluation. The concentration of electrolytes and creatinine is similar to that of plasma, whereas urea and glucose are often slightly elevated. The lactid dehydrogenase content is low but has been reported elevated in malignant cases. Kleist and coworkers have found a significantly elevated content of cholesterol and total lipid in malignant cystic lesions compared to non-malignant in a material of 60 patients, 18 with cystic renal tumors and 42 with benign cysts.

The fluid for cytology should be centrifuged and smeared. Cells are usually few in number and normally of epithelial origin. The background is often clear, but may contain a few erythrocytes and leukocytes.

Contrast medium may be injected at the end of the aspiration procedure in order to perform a radiographic examination of the cyst.

Renal cystic lesions should always be punctured when an abscess is suspected and, if present, treated accordingly (see Chapter 24).

Furthermore, puncture of renal cysts should be performed to relieve pain, if present. However, the cysts often recur. Injection of the oily contrast agent panthopaque has been advocated in order to prevent recurrence because of the sclerosing effect of the medium.

References

Sanders R C: The practical value of diagnostic ultrasound in urology. *J Urol* 1981; 126: 283.

Goldberg B B. Renal tumors. In: Watanabe H, Holm H H, Holmes J H, Goldberg B B, eds. Diagnostic ultrasound in urology and nephrology. Tokyo, New York: Igaku-Shoin, 1981: 31.

Lindgaard D A, Lawson T L. Accuracy of ultrasound in predicting the nature of renal masses. *J Urol* 1979; 122: 724.

Leeham R N, Matzinger A M, Brunelle M, Gray R R, Grosman H. The sonography of renal columna hypertrophy. *J Clin Ultrasound* 1983; 11: 491.

Kristensen J K, Holm H H, Rasmussen S N, Barlebo H. Ultrasonically guided percutaneous puncture of renal masses. *Scand J Urol Nephrol* 1972; 6: (Suppl 15) 49.

Juul N, Torp-Pedersen S, Grønvall S, Holm H H, Koch F, Larsen S. Ultrasonically guided fine needle aspiration biopsy of renal mass lesions. *J Urol* 1985.

Glenthøj A, Partoft S. Percutaneous aspiration of renal angiomyolipoma guided by ultrasound. *Acta Cytologica* 1984; 28: 265.

Murphy J B, Marshall F F. Renal cyst versus tumor: a continuing dilemma. *J Urol* 1980; 123: 566.

Kleist H, Jonsson O, Lundstrøm S, Nauclér J, Nilson A E, Pettersson S. Quantitative lipid analysis in the differential diagnosis of cystic renal lesions. *Br J Urol* 1982; 54: 414.

CHAPTER 13

Ultrasonically guided renal biopsy

Aksel Haubek & Hans Erik Hansen

Several reports dealing with the characteristics of parenchymal renal diseases according to the echo pattern of the diseased kidney have been published. However, present experience indicates that the information obtained is rather unspecific and, to a high degree, operator-dependent. For these reasons a histological biopsy still remains the prerequisite for the classification of medical renal disease and detailed information achieved by electronmicroscopic and immunofluorescence examination makes it even less likely that the percutaneous renal biopsy can be replaced by any non-invasive imaging procedure. Percutaneous aspiration biopsy was introduced more than 30 years ago by Iversen & Brun. Since then the value of renal biopsy has been generally accepted because the information obtained has both prognostic and therapeutic implications and a number of diseases previously ending in renal failure can now be controlled or cured.

Indications

The indications for renal biopsy have changed in the course of time and still vary from center to center. It is gener-ally agreed that clinical and biochemical investigations seldom provide a precise nephrological diagnosis. Since this can usually be provided by a biopsy, percutaneous renal biopsy is indicated in cases of renal failure of uncertain origin.

In patients presenting signs of glomerular lesion with microscopic hematuria and/or proteinuria the indication for biopsy comprises both diagnostic and therapeutic aspects. In minor change nephropathy with the nephrotic syndrome remission can be induced by treatment with steroids or a combination of steroids and cytotoxics. Extracapillary (crescentic) glomerulonephritis is not a biological entity. It occurs as a renal disease without extrarenal lesions, or as part of a systemic disease such as lupus erythematosus disseminatus, polyarteritis nodosa, Wegener's granulomatosis and may be part of Goodpasture's syndrome. Early bioptic verification of this condition is of great importance as its rapidly progressive course to end stage renal failure may be arrested, or at least delayed, if treatment with steroids and cytotoxics is initiated before renal function is severely affected. Acute renal failure due to toxic or toxic/allergic reactions represents an increasing number of cases of acute renal failure and renal biopsy is important in establishing the diagnosis. Finally, the poor or non-functioning renal transplant may constitute the indication for percutaneous biopsy.

Contraindications

Hemorrhagic diathesis, which cannot be corrected, as well as the presence of only one functioning kidney, are regarded as obligatory contraindications. Consequently, information concerning coagulation parameters (as a minimum: hemoglobin, thrombocyte count, bleeding-time, clotting-time and prothrombin index), kidney size and the presence of two functioning kidneys should be obtained before biopsy is performed. Hypertension and fluid retention should be controlled before biopsy, and as bleeding and coagulation abnormalities are elements of the uremic syndrome, severe azotemia should be eliminated by dialysis treatment before performing a renal biopsy. The procedure should be avoided, if possible, in uremic patients with small contracted kidneys, as it may be difficult to obtain tissue and the destruction of the tissue usually is so extensive that no diagnostic information is obtained from the biopsy. The main complication to percutaneous renal biopsy is bleeding – either perirenal or as hematuria. Macroscopic hematuria occurs in about 5–10% of cases, weak or moderate flank pain probably caused by perirenal hematoma in approximately 15%. Severe hemorrhage necessitating blood transfusion occurs and this should be kept in mind when considering renal biopsy, especially in patients with significant cardiovascular disease.

Technique

Different methods may be used for localization of the kidneys and guiding the biopsy needle. A plain film or fluoroscopy has been utilized in combination with the intravenous administration of contrast media. Ultrasound guidance is especially well-suited because the kidneys are visualized regardless of their functional capacity. With static scanning the lower pole of the kidney is localized, the appropriate direction determined and the depth measured. The introduction of dynamic scanning has made the procedure even simpler, but still a two-step technique has been used in many centers – the sonographer localizing the kidney and the nephrologist introducing the needle and performing the biopsy proper. Others have advocated dynamic monitoring of the whole procedure.

The biopsy should be preceded by an ordinary sonographic examination for morphological evaluation of the kidneys. The patient is placed in the prone position and the kidneys are fixed with a pillow under the abdomen. The sample should be obtained from the lower lateral cortex, usually on the left side. The needle is introduced through the puncture device of a dynamic sector scanner. Needles may be chosen according to personal preference and different largebore cutting designs are available.

Results

In our institution renal biopsy has been performed by the nephrologists since the early 'sixties. Since 1982, ultrasonic guidance has been routinely used and the following describes the results from a 19-month period.

A dynamic sector scanner has been used with a puncture attachment. In the majority of cases an Iversen–Roholm needle was used and in some instances a Tru-Cut® needle. When using the former, it was introduced into the capsule of the kidney and the transducer thereafter removed to facilitate the biopsy procedure. In the latter cases the whole procedure was carried out under dynamic monitoring. A total of 46 biopsies was performed in 42 patients. The median age was 47 years with a range of 8–77. The indications were: renal disease of unknown etiology (13 patients), the nephrotic syndrome (6), acute anuria (7), glomerulonephritis (9) and hypertension and impaired renal function of unknown etiology (7 patients).

The success rate was 42 of 46 biopsies. In 2 cases the biopsy was nondiagnostic since only medullary tissue was present and in 2 patients with contracted firm kidneys the biopsy failed. The quality of the biopsies was good. In only 6 instances were less than 5 glomeruli present and in 29 (69%) of the biopsies 10 or more glomeruli were present. This is usually the number considered necessary to quantify the renal affection.

In 8 patients macroscopic hematuria was recognized, and in 4 of these to such an extent that transfusion was necessary. One patient had an unrecognized coagulation defect causing severe and persistent hematuria. When the coagulation defect was treated bladder tamponade developed, necessitating surgery. No patient developed perirenal hematoma requiring treatment.

While the success rate and the quality of the biopsies compares favorably to other reports, the complication rate seems too high. This may reflect the difficulties in introducing a new technique and, according to the literature, a decrease in complications can be expected with growing experience. The nephrologist and the sonographer should be equally acquainted with the technique and indications and contraindications should be discussed in close cooperation.

References

Iversen P, Brun C. Aspiration biopsy of the kidney. *Am J Med* 1951; 11: 324.

Brun C, Olsen S. Atlas of renal biopsy. Philadelphia, London, Toronto: Munksgaard/W.B. Saunders, 1981: 9.

Pollack H M, Goldberg B B, Kellerman E. Ultrasonically guided renal biopsy. *Arch Intern Med* 1978; 138 (3): 355.

Mets T, Lameire N, Matthys E, Afschrift M. Sonically guided renal biopsy. *JCU* 1979; 7: 190.

Backman U, Lindgren P G. Percutaneous renal biopsy with real-time ultrasonography. *Scand J Urol Nephrol* 1982; 16: 65.

Ultrasonically guided prostatic biopsy

Niels Juul, Søren Torp-Pedersen & Maxwell Sehested

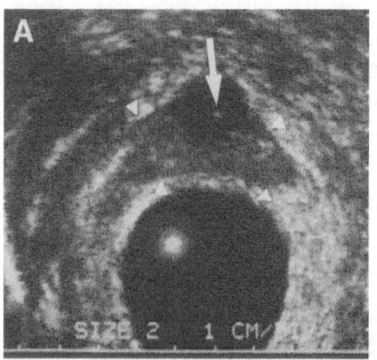

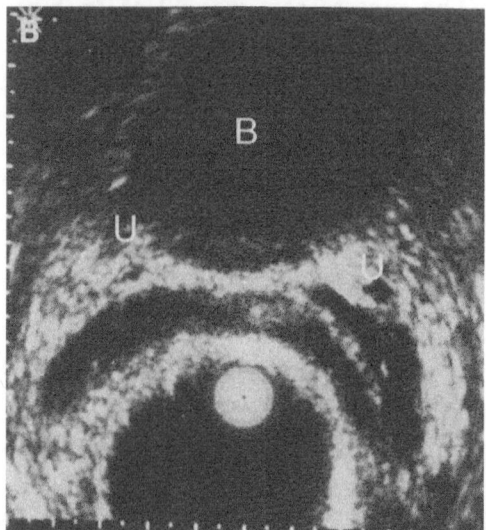

The urinary bladder, the seminal vesicles and the prostate can be visualized by abdominal ultrasonic scanning. However, when detailed information about the prostate and its surroundings is required this approach is inferior to the transrectal.

The first transrectal scanner was developed by Wild & Reid in 1957. Although they were only able to visualize part of the rectal wall, their work formed the basis for more recent scanners of the rotating type. The first tomograms of the prostate for clinical use were obtained by Watanabe. The patient was sitting in a specially designed chair with the rotating scanner in the rectum covered by a water-filled balloon. The technique has been further developed by others using the left lateral decubitus position.

On the ultrasound scan the normal prostate appears triangular, well delineated anterior to the rectum. The periurethral gland can be outlined as an echopoor structure in the anterior part of the prostate (Fig. 14-1A). By advancing the scanner further into the rectum the seminal vesicles can be visualized echofree and "horseshoe-shaped" between the rectum and the bladder (Fig. 14-1B).

In the case of benign prostatic hy-

Fig. 14-1. Normal prostate and seminal vesicles

A: Transrectal scan using rotating probe (Brüel & Kjær, Nærum, Denmark) showing a normal prostate. The periurethral gland (arrow), as well as the total prostate (arrowheads), is clearly outlined.

B: The normal seminal vesicles are seen as an echo-free "horseshoe" between the bladder (B) and the rectum.

U: Ureters.

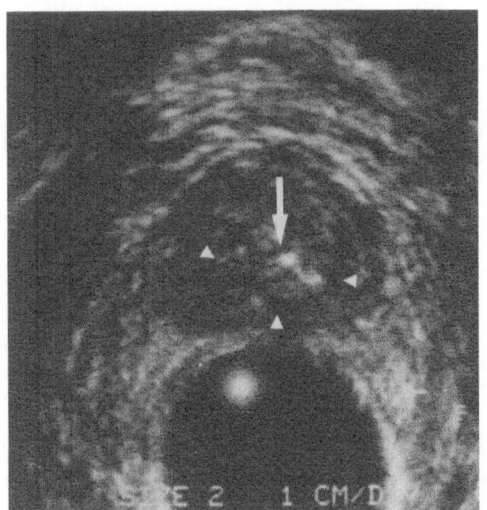

Fig. 14-2. Benign prostatic hypertrophy
The gland is large and rounded. The adenoma is clearly outlined (arrowheads). Catheter in urethra (arrow).

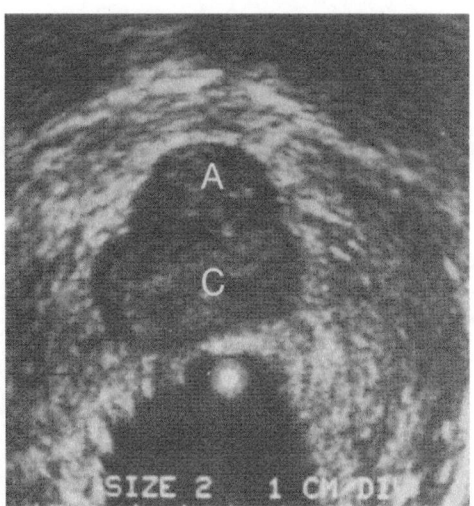

Fig. 14-3. Intracapsular prostatic cancer
The posterior lobe with the carcinoma (C) is "pushing" the adenoma (A) anteriorly.

pertrophy the prostate is seen enlarged, rounded with a centrally placed adenoma and a compressed surgical capsule (Fig. 14-2).

In early prostatic cancer, echo-rich as well as echo-poor areas may be seen, especially in the posterior-lateral part of the gland. Furthermore, the architecture of the gland may be disturbed (Fig. 14-3). Although these findings may suggest malignancy, they may also be seen in prostatitis or even with a normal prostate. Only when based upon visualization of periprostatic growth or capsular involvement is the diagnosis of prostatic cancer reliable (Fig. 14–4).

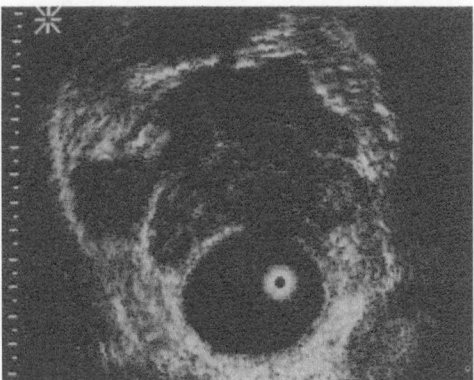

Fig. 14-4. Advanced prostatic cancer
Extensive periprostatic tumor growth is noted.

Biopsy technique and experience

A method for ultrasonically guided precise needle placement in the prostate and seminal vesicles has been described by Holm & Gammelgaard. With the patient in the lithotomy position and the transrectal scanner placed in an external fixture, a biopsy needle can, via an attachable needle steering device, be introduced transperineally, parallel to the scanner axis into the prostate (Fig. 14-5A). The parallel canals in the needle steering device correspond to a puncture line on the monitor. The scanner is fixed with the puncture line crossing the suspect area and the needle is intro-

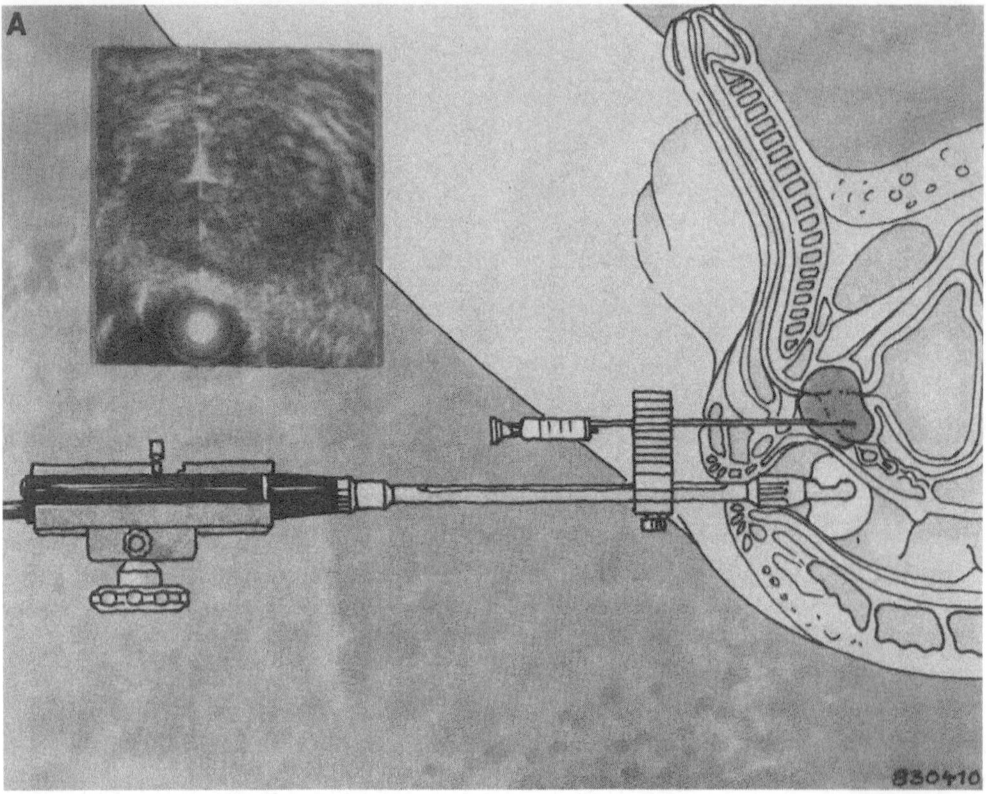

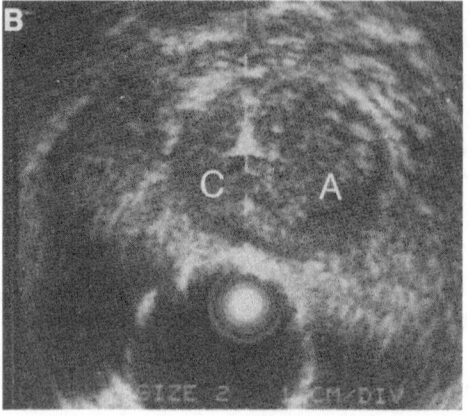

Fig. 14-5. The principle of ultrasonically guided transperineal biopsy
A: When the needle crosses the image plane a bright echo appears on the monitor.
B: Transperineal biopsy from thickened right lateral lobe of the prostate (C). The needle tip echo ensures that the biopsy is taken from the correct place. A: Adenoma.

duced through the corresponding canal. Thus any predetermined area in the prostate can be punctured and correct needle placement is verified by the appearance of a bright echo on the screen (Fig. 14-5B).

Hastak et al. performed transrectal ultrasound scanning and ultrasonically guided transperineal biopsy with a Tru-Cut® needle in 16 consecutive patients admitted with clinical suspicion of prostatic cancer. The final diagnosis obtained by transurethral resection was cancer in 8 cases, benign prostatic hypertrophy in 5 cases and 3 cases of prostatitis. The correct diagnosis was established in all cases from the biopsy specimens. However, the diagnosis obtained by the ultrasound scanning alone was incorrect in 5 cases, including 4 false

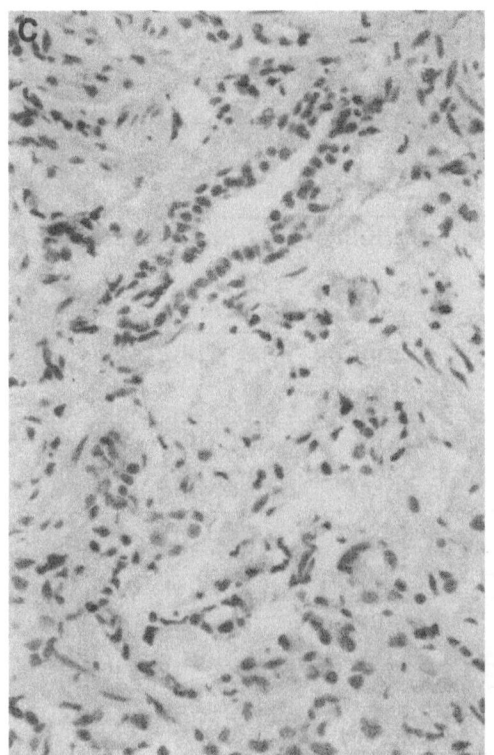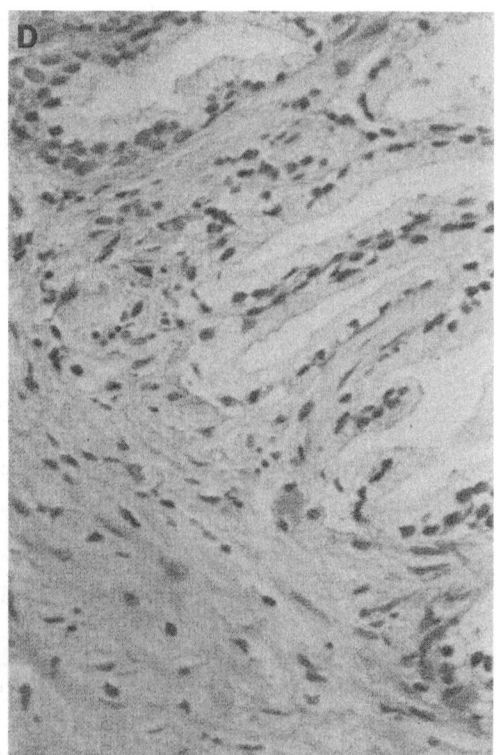

C: Histological specimen obtained from the suspected area showing adenocarcinoma.
D: Histological specimen obtained from the adenoma showing benign glandular hyperplasia.

positive cancer diagnoses. They concluded that ultrasonically guided transperineal biopsy is a reliable tool in the diagnosis of prostatic malignancy.

Transperineal biopsy guided by a linear array probe has been described by Fornage et al. and by Rifkin et al.

Biopsy needles

A fine needle aspiration biopsy yields material for cytological evaluation. However, the interpretation is difficult and the method has not gained wide acceptance. Most prostatic biopsies are therefore carried out with large bore cutting needles. The discomfort of a large bore biopsy needle is, however, by no means negligible and it is not suited for out-patient use. A preliminary study of prostatic biopsies with a 21 gauge modified Menghini needle (Surecut®) is promising and it may be used as an alternative.

With this Surecut® needle it is possible to obtain tissue cores 0.6 mm in diameter with a length of up to 3 cm. The tissue core follows routine histological processing and regular histological diagnoses are obtained. It is therefore possible to combine the advantages of a large bore biopsy with those of a fine needle biopsy – i.e., material for histological evaluation is obtained with minimal discomfort to the patient.

In 14 consecutive cases, Torp-Peder-

Table 14-1. Results of ultrasonically guided transperineal Surecut® biopsies

	Palpation	Ultrasound	Surecut®	Verification – T stage
1.	benign hypertrophy	adenoma with echo-poor regions. Biopsy from echo-poor regions.	no malignancy	no verification
2.	benign hypertrophy	adenoma with echo-poor regions. Biopsy from echo-poor regions.	no malignancy	no verification
3.	benign hypertrophy	2 adenomas. Biopsy from smallest adenoma asymmetrically located.	no malignancy	TUR-P no malignancy
4.	benign hypertrophy and thickening to the left	benign hypertrophy Biopsy from left side of posterior lobe	no malignancy	TUR-P no malignancy
5.	prostatic cancer	thickened posterior lobe	prostatitis	no verification
6.	prostatic cancer	normal ultrasound. Biopsy from posterior lobe	no malignancy	Tru-cut® biopsy: no malignancy
7.	prostatic cancer	left side lobe very echo-rich	adenocarcinoma moderately differentiated	Bony metastases – no TUR-P T_2
8.	prostatic cancer	asymmetrical gland with thickening in the left side	insufficient material	TUR-P: adenocarcinoma moderately differentiated. Bony metastases T_2
9.	prostatic cancer	posterior lobe slightly thickened to the left	adenocarcinoma poorly differentiated	TUR-P: adenocarcinoma moderately and poorly differentiated. Bony metastases T_2
10.	prostatic cancer	adenoma asymmetrically located. Biopsy from thickened posterior lobe	adenocarcinoma highly differentiated	TUR-P: adenocarcinoma highly differentiated. Bony metastases T_2
11.	prostatic cancer	thickning of posterior lobe to the left	adenocarcinoma moderately differentiated	no TUR-P: Bony metastases T_2
12.	prostatic cancer	irregular gland with capsular defect	adenocarcinoma poorly differentiated	no verification T_3
13.	Hard prostate. Control after radioactive seeds implantation	small prostate with seeds	fibrosis	no verification
14.	Hard prostate. Control after radioactive seeds implantation	small prostate with seeds	fibrosis	no verification

sen and coworkers performed ultra-sonically guided transperineal biopsies of the prostate using this needle. The biopsies were carried out as previously described with a slight modification. To facilitate the entry of the fine needle it was introduced through an 18 gauge lumbar needle which only penetrated the skin. The findings at palpation, ul-trasound study and biopsy are shown in Table 14-1. There was 1 case of insuffi-cient material giving a retrieval rate of 93%. In this small material there were no false positives and no false negatives. It is to be expected that false negatives will occur as they do in all biopsy ma-terials because of sampling errors. The advantage of this method, however, is that histological material is obtained with a fine needle. Furthermore the method enables the pathologist to grade the tumor and therapy can be planned accordingly.

References

Wild J J, Reid J M. Progress in technique of soft tissue examination by 15 MC pulsed ultrasound. In: Kelly E, ed. Ultrasound in Biology and Medicine. Washington: AIBC 1957: 30.

Watanabe H, Kato H, Kato T, Tanaka M, Terasawa Y. Diagnostic application of the ultrasonotomography to the prostate. *Jap J Urol* 1968; 59: 273.

Hallemans E; Declercq G, Denis L. trans-rectal ultrasonotomography. *Eur Urol* 1977; 3: 37.

Gammelgaard J, Holm H H. Transurethral and transrectal ultrasonic scanning in urology. *J Urol* 1980; 124: 863.

Holm H H; Gammelgaard J: Ultrasonically guided precise needle placement in the prostate and seminal vesicles. *J Urol* 1981; 125: 385.

Hastak S M, Gammelgaard J, Holm H H. Ultrasonically guided transperineal bi-opsy in the diagnosis of prostatic carci-noma. *J Urol* 1982; 128: 69.

Rifkin M D, Kurtz A B, Goldberg B B. Prostate biopsy utilizing transrectal ultra-sound guidance. *J Ultrasound Med* 1983; 2: 165.

Fornage B D; Touche D H, Deglaire M, Fa-roux M C, Simantos A. Real-time ultra-sound-guided prostatic biopsy using a new transrectal linear-array probe. *Radi-ology* 1983; 146: 547.

O'Donahue E P N: Biopsies. In: Chisholm C D. ed. *Urology*. London: Medical Books Ltd, 1980.

Torp-Pedersen S, Juul N, Sehested M, Ras-mussen F. Prostatic biopsy with the 21 gauge Surecut® needle. *Br J Urol.* 1984; 57: 43.

CHAPTER 15

Ultrasonically guided puncture of pancreatic mass lesions

S. Hancke, H. H. Holm & F. Koch

Ultrasound scanning is, in many institutions, the method of choice for visualization of the pancreas and for the diagnosis of various pancreatic diseases.

Ten years ago, the first ultrasonically guided fine needle aspiration biopsy from the pancreas was performed (Hancke et al. 1975), and in many centers the procedure is now considered routine in the evaluation of patients with suspected pancreatic malignancy.

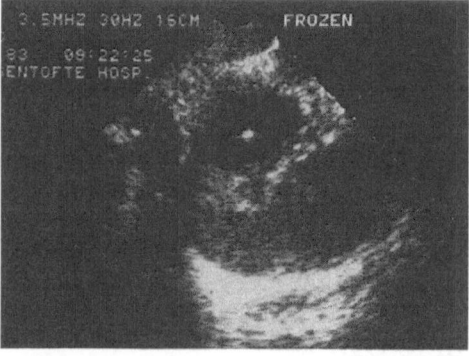

Fig. 15-2.
Dynamic sector scan demonstrating a 3 cm cystic mass lesion in the tail of the pancreas. The echo from the tip of the puncture needle is seen in the center of the cyst.

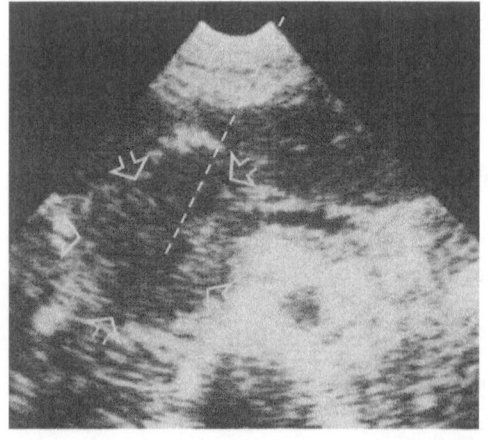

Fig. 15-1. Solid mass lesion of the pancreas
Dynamic transverse sector scan of the upper abdomen demonstrating a 3 cm solid mass lesion in the head of the pancreas (arrows). Dotted line indicates the route for the fine needle introduced obliquely into the sound field.

When it is impossible from the ultrasound examination and the clinical setting to differentiate between a malignant and benign mass lesion of the pancreas, percutaneous fine needle biopsy should be performed (Fig. 15-1).

The first ultrasonically guided puncture of a cystic pancreatic lesion was performed nine years ago (Hancke and Pedersen, 1976), and percutaneous aspiration is now a routine in the diagnosis and treatment of pancreatic cysts and abscesses (Fig. 15-2).

Solid mass lesions

Method

As with routine scanning of the pancreas, the patient should be fasted for about 8 hours before the puncture. No further preparation is needed except bleeding and coagulation studies in patients with suspected hemorrhagic diathesis. Jaundice or portal hypertension are *not* contraindications.

Although static scanning may be used to guide the puncture, dynamic scanning, especially dynamic sector scanning, is preferable since it provides the best identification of the needle during its insertion.

In most cases a guide needle (1.2 mm od) is first inserted through the abdominal wall. This guide needle allows several passes of the fine needle (0.6 mm o.d.). In cases of large tumors of the pancreas, where the mass is located close to the anterior abdominal wall, a fine needle with a stylet may be introduced directly through the skin into the lesion. Two to 4 passes are made routinely. Vacuum is applied from a 10 ml syringe mounted in a one-handed aspiration grip.

In our institutions the procedure is often performed on an out-patient basis. In such cases the patient rests in an observation room for about 15 minutes after the biopsy.

Material and results

At the Gentofte and Herlev Hospitals, Copenhagen, ultrasonically guided fine needle biopsy from the pancreas has been performed in 203 patients during a 10-year period (Table 15-1).

Table 15-1. Ultrasonically guided fine needle aspiration biopsy of the pancreas

| | | Cytology | | |
		+	−	Insuff.
Final Diagnosis	Malignant mass	126	90	30	6
	Benign condition	77	2	68	7
	Total	203	92	98	13

The series comprises 126 patients with a malignant mass lesion of the pancreas or a malignant periampullary tumor. In 90 of these 126 patients malignant cells were demonstrated cytologically. Thirty patients had a false negative cytological diagnosis, giving a false negative rate of 24%.

Seventy-seven patients were without malignancy, most of whom had chronic pancreatitis. In 2 of these, false positive cytological diagnoses were obtained. Both had chronic pancreatitis. At repeat biopsy in one of these patients some weeks later a correct negative diagnosis for cancer was obtained.

In 13 cases (6.4%) the aspirate was insufficient for a cytological diagnosis.

The predictive value of the positive cytological diagnosis (pVpos) is 98% while the predictive value of the negative diagnosis (pVneg) is 69% in the present series.

No serious complication occurred among the 203 patients.

Cystic mass lesions

Method

As with fine needle puncture the patient should be fasted, but no further preparation is needed. Dynamic scanners, linear array or sector scanners are most useful in the guidance of the puncture needle. Depending on the size of the cyst and on the amount of debris, a 0.9 mm or 1.2 mm outer diameter needle is used. The puncture site is in the epigastrium or laterally at an intercostal space. Quantitative aspiration of the cystic cavity is usually attempted.

In cases of larger cysts which are close to the abdominal wall a percutaneous catheter can be introduced into the cystic cavity in order to drain the cyst temporarly or definitively.

Pancreatic cyst aspirations are, in our institutions, often carried out on an outpatient basis.

Material and results

In the past 9 years ultrasonically guided puncture of cystic lesions in the pancreatic region has, in our institutions, been performed whenever disclosed by ultrasound.

In a series of 100 patients 33 were punctured in an acute stage, 57 patients were punctured electively during a period of chronic pancreatic disease and 10 patients were punctured in order to diagnose an incidentally demonstrated cyst in the pancreatic region.

In the 100 patients 122 cysts were punctured. Because of recurrence a total of 194 punctures were performed. The cysts were located with equal incidence in the head, body and tail of the pancreas.

The size of the cysts punctured varied from 2 to 20 cm in diameter. 28 cysts were smaller than 4 cm in diameter.

In most cases the cyst fluid aspirated was cloudy or brownish. In 3 incidences the fluid was bloody. In 2 of the cases because of arterial bleeding into the cyst

prior to the puncture. In one case an aneurism of the hepatic artery, misdiagnosed as a cyst, was punctured. Pus was aspirated in 13 cases while tumorcells were isolated from the aspirated fluid in one patient with a cystadenocarcinoma. One other patient had also a cystadenocarcinoma, but tumorcells were not demonstrated in the cystic fluid from this patient. However, a malignant diagnosis was established preoperatively in this patient by fine needle aspiration from a solid portion of the lesion.

Relief of pain was observed regularly after the puncture, the duration being from a few days to permanent.

Intracystic catheter placement was performed in 16 patients. 13 of these were later operated upon because of recurrence of the cysts, while 3 patients recovered without operation. It should be noted, that 2 of these 3 patients were drained for pancreatic abscesses.

Spontaneous regression of pancreatic pseudocysts is described in the literature as varying from 5 to 8%. In 8 out of 102 cysts punctured in our series, no recurrence was recorded within a 12 months observation period. In 13,7% the cysts did not recur within 2 months (14/102).

Possible complications to percutaneous puncture are bleeding, infection or fistula formation.

In our series of 194 punctures we have experienced only one 200 ml intraperitoneal hematoma which was found at surgery 3 months after the puncture.

In no instances infection or fistulae formation were recorded.

As described above a misdiagnosed aneurism was punctured. 3 days later the patient died because of rupture of the aneurism. However, it is hard to say whether the needle puncture was fully responsible for the accident.

Discussion

Ultrasound scanning in combination with fine needle aspiration biopsy is now a well established method in the evaluation of patients with suspected pancreatic cancer.

With modern scanning equipment offering very precise needle placement, an increasingly high diagnostic accuracy should be expected. However, the figures have not improved during the last 10 years. The reason for this might be that liberal use of ultrasound scanning in patients with only vague symptoms now discloses pancreatic lesions at an early (and small) stage.

In our series, 2 false positive cytological diagnoses occurred. However, false positive diagnoses are not reported elsewhere in the literature. Thus, from reports of hundreds of patients the rate of false positives is very low, less than 1%. The false negative rate, however, is rather high, varying from 12–33% as reported in the literature. Therefore pancreatic cancer cannot be excluded by the technique described, but when malignant cells are demonstrated cancer is accurately predicted.

It should be stressed that cytological evaluation demands great care and experience. A close collaboration with the cytologist is of the utmost importance.

The method has proved safe throughout the years. In our institutions no serious complication has been recorded. In some cases an increased level of amylase in blood or urine was observed for a few days after the puncture, but no instances of bleeding, infection or fistula forma-

tion occurred. In the literature, cases of acute pancreatitis in connection with fine needle puncture have been reported. A single case of fatal necrotizing pancreatitis followed by fine needle biopsy was reported in 1981 (Evans et al.). In this case the puncture was performed under flouroscopic control by 4 transabdominal needle passes in a patient with suspected enlargement of a non-malignant pancreas. A case of hemorrhagic pancreatitis has also been described (Dzieniszewski et al.). Extra caution should be exercised when no clearly defined mass lesion is visible within the pancreas. To our knowledge seeding of tumor cells by fine needle puncture guided by ultrasound scanning has not been observed.

Puncture of cystic mass lesions of the pancreas is diagnostic as well as therapeutic.

In the acutely ill patient it is possible with an ultrasonically guided puncture of a cystic mass to differentiate a simple pseudocyst of the pancreas from an abscess in the region. In patients with chronic pancreatic disease a percutaneous puncture can verify the presence of a pseudocyst of the pancreas, when the amylase in the cystfluid is elevated. With injection of contrastmaterial a possible communication to the pancreatic duct can be demonstrated. Through cytological study of the fluid aspirated, a possible cystadenocarcinoma can be disclosed.

Reviewing the therapeutic purposes it is obvious that in the acutely ill patient a cystic lesion can be drained by needle puncture one or several times. Thereby threatening rupture of a pancreatic cyst can be avoided. Through repeated aspirations the cystmembrane matures facilitating elective surgical treatment. In patients with chronic pancreatitis and pseudocysts a single or repeated puncture with quantitative aspiration can provide a variable degree of pain relief and in cases of obstruction caused by pressure from a pancreatic cyst, puncture can relieve biliary or gastric outlet obstruction, in some cases temporary, in other cases permanent.

Finally definitive therapy can be the result of an ultrasonically guided cyst puncture. This is the case when the cyst does not recur although the basic disease is not cured. It may also be the case if the patient has a permanent relief of pain after aspiration – although the cyst is still present.

Conclusion

Ultrasonically guided fine needle biopsy of the pancreas is most valuable in the demonstration of cancer and less valuable for the exclusion of malignancy. The method, which is simple, rapid and accurate and carries a very low risk, should be performed whenever a pancreatic mass is detected. When malignancy is demonstrated preoperatively the surgeon may proceed confidently with a palliative or radical operation. If the patient is unfit for pancreatic resection and/or does not have gastric outlet obstruction or cholestasis, exploratory laparotomy can be avoided.

Pancreatic cyst puncture should be performed in the acute as well as in the chronic stage of the disease to obtain a diagnosis of pseudocysts, abscesses, cyst communication to the pancreatic duct and malignant cysts. It should also be performed therapeutically for threatening rupture, cystwall maturation, relief of pain or obstruction and in an attempt to avoid surgery.

Further studies should be planned to evaluate the use of temporary drainage by catheterization. This kind of treatment might be more useful than conventional surgical treatment.

References

Hancke S, Holm H H, Koch F. Ultrasonically guided percutaneous fine needle biopsy of the pancreas. *Surg Gynecol Obstet* 1975; 140: 361.

Hancke S, Pedersen J F. Percutaneous puncture of pancreatic cysts guided by ultrasound. *Surg. Gynecol. Obstet.* 1976; 142: 551.

Braun B, Dormeyer H H. Ultrasonically guided fine needle aspiration biopsy of hepatic and pancreatic space-occupying lesions and percutaneous abscess drainage. *Klin Wochenschr* 1981; 59: 707.

Hancke S. Ultrasound in the diagnosis of pancreatic cancer. Scanning and percutaneous fine needle biopsy. Stockholm: Almqvist & Wiksell International, 1981.

Hovdenak N, Lees W R, Pereira J, Beilby J O, Cotton P B. Ultrasound guided percutaneous fine needle aspiration cytology in pancreatic cancer. *Br Med J Clin Res* 1982; 285: 1183.

Labadie M, Descos L, Berger F. Cytodiagnosis by needle aspiration biopsy in pancreatic diseases, a preliminary study. *Semin Hop* Paris 1981; 57: 261.

Schwerk W B, Schmitz-Moormann P. Ultrasonically guided percutaneous transperitoneal fine needle biopsy of pancreatic tumour. *Dtsch Med Wochenschr* 1980; 105: 1019.

An-Foraker S H, Fong-Mui K K. Cytodiagnosis of lesions of the pancreas and related organs. *Acta Cytol* 1982; 26: 814.

Bret P M, Fond A, Bretagnolle M, Barrel F, Labadie M. Percutaneous fine needle biopsy (P.F.N.B.) of intra-abdominal lesions. *Eur J Radiol* 1982; 2: 322.

Ferruci J T, Wittenberg J, Mueller P R, et al. Diagnosis of abdominal malignancy by radiological fine needle aspiration biopsy. *AJR* 1980; 134: 323.

Ohto M, Karasawa E, Tsuchiya Y, et al. Ultrasonically guided percutaneous contrast medium injection and aspiration biopsy using a real time puncture transducer. *Radiology* 1980; 136: 171.

Porter B, Karp W, Forsberg L. Percutaneous cytodiagnosis of abdominal masses by ultrasound guided fine needle aspiration biopsy. *Acta Radiol Diagn* 1981; 22: 663.

Yamanaka T, Kimura K. Biopsy of the pancreas under the guidance of ultrasonic imaging technics and its effectiveness in the diagnosis of pancreatic neoplasms. *Nippon Rinsho* 1980; 38: 127.

Evans W K, Ho C-S, McLoughlin M J, Tao L-C. Fatal necrotizing pancreatitis following fine-needle aspiration biopsy of the pancreas. *Radiology* 1981; 141: 61.

Dzieniszewski G P, Neher M, Linhart P, Frank K. Necrotizing pancreatitis after ultrasonically guided fine needle aspiration biopsy. *Dtsch Med Wochenschr* 1982; 107: 1438.

Andersen B N, Hancke S, Damgaard S A, Schmidt A. The diagnosis of pancreatic cysts by endoscopic retrograde pancreatography and ultrasonic scanning. *Annals Surgery* 1977; 185: 286.

MacErlean D P, Bryan P J, Murphy J J. Pancreatic pseudocyst: Management by ultrasonically guided aspiration. *Gastrointest Radiol* 1980; 5: 255.

Schwerk W B. Ultrasonically guided percutaneous puncture and analysis of aspirated material of cystic pancreatic lesions. *Digestion* 1981; 21: 192.

Barkin J S, Smith F R, Pereiras R Jr., Isikoff M, Levi J, Livingstone A, Hill M, Rogers A I. Therapeutic percutaneous aspiration of pancreatic pseudocysts. *Digestive Dis Sciences* 1981; 26: 585.

Karlsson K B, Martin E C, Fancuchen E I, Mattern R F, Schultz R W, Casarella W J. Percutaneous drainage of pancreatic pseudocysts and abscesses. *Radiology* 1982; 142: 619.

105

Ultrasonically guided percutaneous pancreatography

Kenichi Takayasu, Masatoshi Makuuchi & Munemasa Ryu

Percutaneous pancreatography under ultrasonic guidance was first described by Cooperberg et al. in 1979. With the development of real-time ultrasonography, direct puncture of the pancreatic duct has become much easier and safer, as it has for other vessels and organs. Using a mechanical or electronic linear array real-time scanner, we began this procedure for percutaneous pancreatography in various pancreatic diseases in June, 1978, and have now accumulated sufficient experience with which to evaluate its clinical significance.

Instruments and procedure

Fig. 16-1 shows two kinds of commercially available transducer for puncture. Fig. 16-1A shows a 3.5 MHz linear array, electronic real-time scanner which has a center canal shaped like an inverted triangle (Toshiba Medical Co., Tokyo), and a 3.0 MHz mechanical real-time sector scanner with an adaptor on its lateral side (Fig. 16-1B) (Aloka Medical Co., Tokyo). An 18 cm long, 22 gauge thin Chiba needle and a 12 cm long guide needle with 0.8 mm inner diameter and 1.0 mm outer diameter were used for puncture. As preoperative preparation, fasting on examination day and pentazocine 15–30 mg i.m. before examination were required. After scanning the pancreas with the puncture transducer, a suitable position for puncture is decided. If the left lobe of the liver or the stomach lie in front of the pancreas, the puncture needle is passed through these organs. The guide needle is first inserted after local anesthesia of the skin and then the Chiba needle is advanced through the guide needle under echographic monitoring. While detaching the puncture transducer, a 15 cm long extension tube is attached to the tail of the Chiba needle in order to prevent any external force from being transmitted to the needle. After aspiration of pancreatic juice, about the same amount of 65% angiografin with antibiotics (gentamicin) is slowly injected under X-ray monitoring. Sometimes no fluid is aspirated, but contrast injection is nevertheless done. After the examination, as much as possible of the injected contrast medium is removed in order not to aggravate the pancreatitis combined with carcinoma. Aspirated fluid is analyzed for amylase concentration, cultured for bacteria, and cytological examination is carried out using Papanicolaou and May-Giemsa stains.

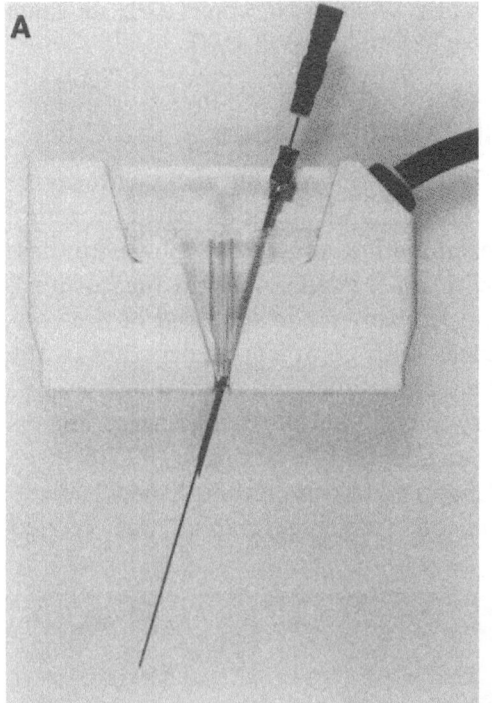
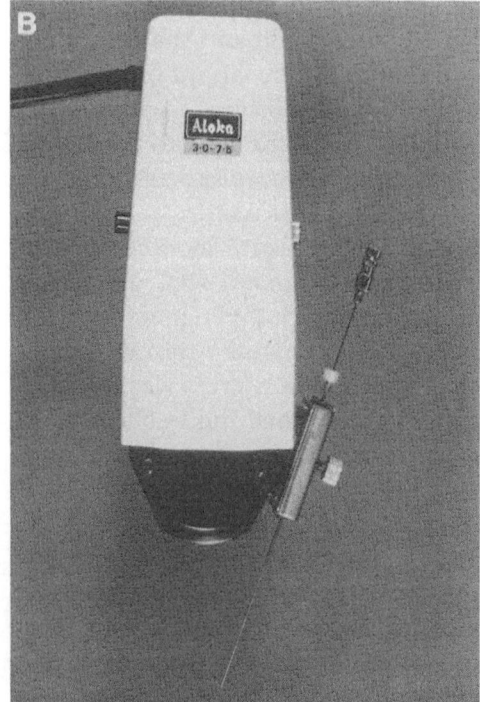

Fig. 16-1. Two kinds of transducer for puncture
A). A linear array, electronic real-time scanner with an inverted triangle-shaped puncture canal, and a removable adaptor. B). A mechanical sector scanner with an adaptor for puncture on the lateral side.

Material

Percutaneous pancreatography was performed in a total of 41 patients, of whom 23 had cancer (21 had carcinoma of the head of the pancreas and 2 carcinoma of the body), 11 had chronic pancreatitis and 7 had ampullary cancer. In all but 1 with carcinoma, histological diagnosis was obtained. There were 27 males and 14 females, age range from 41 to 82 years. Endoscopic retrograde pancreatography (ERP) failed in 19 cases due to gastrectomy, pancreatic divism and/ or technical failure.

Results

Percutaneous pancreatography was suc-

cessful in 38 out of 41 patients (92.7%) with a mean of 1.7 punctures (Table 16-1). In 2 of 3 patients in whom puncture failed, the caliber of the pancreatic duct was less than 5 mm, and the failure was thought to be due to a technical fault in

Table 16-1. Results and complications

Results	
1) Success rate	38/41 (92.7%)
2) Number of punctures (mean)	1.7 times
Complications	
1) Transient abdominal pain	2/41 (4.9%)
2) Findings at operation	2/32 (6.3%)
small hematoma on pancreas	1
small amount of fluid in the lesser omentum	1

These complications did not occur in the same patients.

one case. For a 5 mm pancreatic duct, 3 attempts were made and the 3rd puncture succeeded. The largest caliber pancreatic duct was 18 mm. In all patients with dilatation of the duct to more than 10 mm, puncture was successful at the first attempt. There were no major complications. Transient abdominal pain not requiring treatment occurred during the examination in 2 patients (4.9%), a small hematoma about 5 mm in size on the surface of the pancreas was found in 1 patient and a small fluid collection in the lesser omentum in another at surgery (Table 16-1). In 3 patients ERP failed to opacify the pancreatic duct, even though cannulation was successful, but pancreatography clearly visualized the duct. Moreover, percutaneous pancreatography helped in reaching a differential diagnosis of chronic pancreatitis with inflammatory mass formation from carcinoma in 2 cases (Fig. 16-5). Table 16-2 gives the analysis of the aspirated pancreatic fluid. Cytological examination disclosed malignant cells in only 2 of 20 patients with cancer. Thus, cytology of pancreatic fluid seems to have little diagnostic significance for cancer. Bacterial culture was positive in 2 of 10 patients, perhaps as a result of preceding endoscopic retrograde pancreatography. The concentration of amylase was higher than 10,000 IU/L in about 65% of all cases with pancreatic cancer, whereas, in about 27% its value was less than 1,000 IU/L.

Illustrative cases

Case 1. A 63-year-old male with severe jaundice and itching. An ultrasound examination revealed an echo-poor mass with an irregular margin measuring 3 cm in diameter in the head of the pancreas with complete obstruction of the pancreatic duct and dilation of the biliary ducts. Therefore, after percutaneous transhepatic bile drainage (PTBD) (see Chapter 8) was accomplished, percutaneous pancreatography was carried

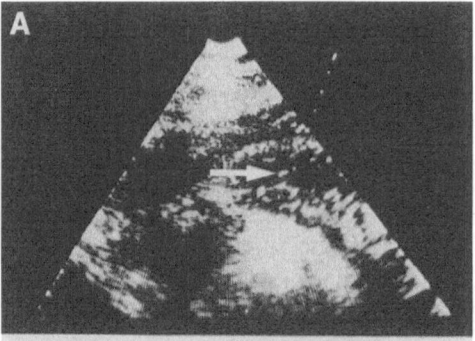
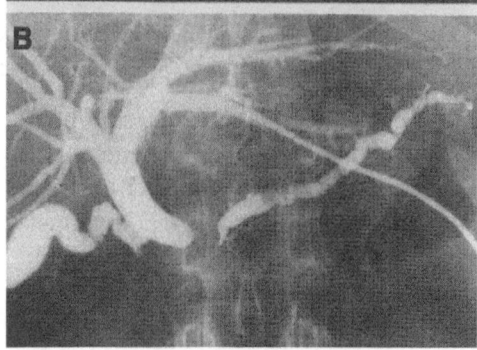

Fig. 16-2. Case 1. **Carcinoma of the Pancreas**
A: Carcinoma of the head of the pancreas. Under echographic monitoring, a needle was inserted into a dilated pancreatic duct (Needle tip is visualized, indicated by arrow). B: Percutaneous pancreatography combined with tube cholangiography revealed complete obstruction of both ducts in the head of the pancreas.

Table 16-2. Analysis of aspirated pancreatic juice

1) Cytology: positive	2/20 (10.0%)
2) Bacterial culture: positive	2/10 (20.0%)
3) Concentration of amylase (IU/1)	
$\leq 10^3$	3/11 (27.3%)
$10^3 — 10^4$	1/11
$> 10^4$	7/11 (63.6%)

out (Fig. 16-2A). Complete obstruction of both ducts was seen and the exact location and extent of tumor invasion was delineated (Fig. 16-2B). Aspiration of the pancreatic fluid was performed; amylase was 32,400 IU/L, carcino-embryogenic antigen 285 ng/ml and cytological examination of the aspiration fluid was positive. The diagnosis was obviously carcinoma of the head of the pancreas. This was confirmed and a small fluid collection in the lesser omentum recognized at operation.

Case 2. A 63-year-old male presented with abdominal pain. Ultrasound examination disclosed a markedly dilated

pancreatic duct, many intraductal strong echoes without an acoustic shadow and moderately dilated biliary ducts. ERP failed because the proximal part of the pancreatic duct was obstructed by numerous stones. Therefore, percutaneous pancreatography was carried out. It revealed filling defects in the dilated pancreatic duct and stenosis of the choledochus in the pancreatic portion, which was demonstrated by PTC (Fig. 16-3). Chronic pancreatitis with many intraductal stones was diagnosed.

Case 3. A 43-year-old male with a long history of abdominal pain. At the outpatient clinic, a sonolucent mass with a

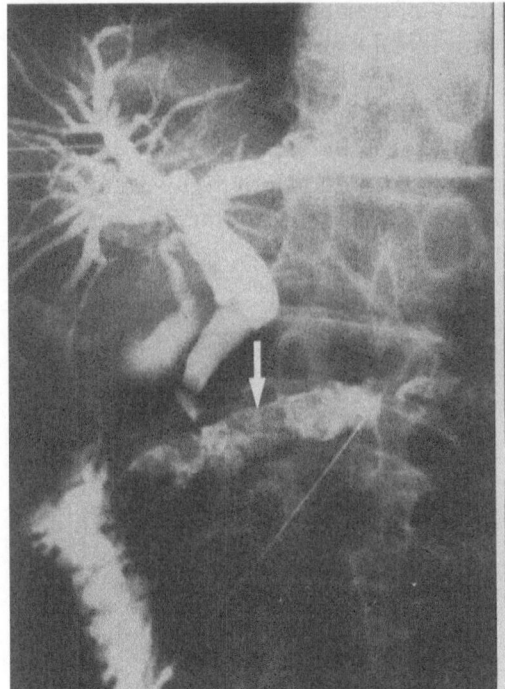
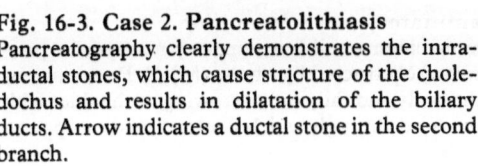

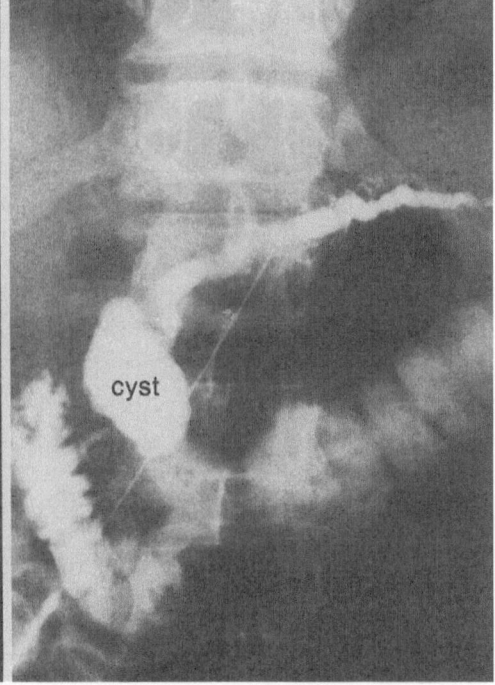

Fig. 16-3. Case 2. Pancreatolithiasis
Pancreatography clearly demonstrates the intraductal stones, which cause stricture of the choledochus and results in dilatation of the biliary ducts. Arrow indicates a ductal stone in the second branch.

Fig. 16-4. Case 3. Chronic pancreatitis with a pseudocyst
Pancreatography disclosed a solitary cyst in the head of the pancreas, communicating with a dilated and tortuous pancreatic duct.

slightly irregular margin and a mildly dilated pancreatic duct were demonstrated by ultrasound. This mass was first thought to be a cancer of the head of the pancreas, but the final diagnosis based on the percutaneous pancreatography (Fig. 16-4) was chronic pancreatitis with a pseudocyst communicating with the pancreatic duct, and surgery was called off.

Case 4. A 51-year-old male with abdominal pain. Fig. 16-5A shows a sonolucent mass measuring 3 cm with complete obstruction of the dilated pancreatic duct. It was strongly suspected to be a cancer of the head of the pancreas. However, percutaneous pancreatography demonstrated communication of the dilated pancreatic duct with the duodenum, tapered narrowing, but mainly displacement of the duct and well-visualized branches around it (Fig. 16-5B); thus a diagnosis of chronic pancreatitis accompanied with inflammatory mass formation was made. Because of the severe abdominal pain, the patient underwent pancreato-jejunostomy, and was still alive after 2 years. No cancer tissue was found histologically.

Case 5. A 71-year-old male who presented with jaundice. A combination study of percutaneous pancreatography and PTBD (Fig. 16-6) revealed complete obstruction of both ducts. From these findings, a diagnosis of periampullary cancer of the pancreas was made. Hypotonic duodenography later gave more details corroborating the diagnosis.

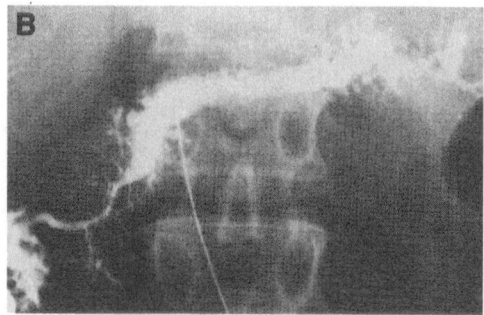

Fig. 16-5. Case 4. Chronic pancreatitis with inflammatory mass formation
A: By sonography alone, differential diagnosis between carcinoma and inflammatory mass of the pancreas is not possible. Note the dilated pancreatic duct (arrow) obstructed by a mass (M). B: Percutaneous pancreatography revealed a dilated pancreatic duct in the body and tail, and narrowing of the duct with a smooth ductal wall and well-visualized branches in the head through which contrast medium passed to the duodenum.

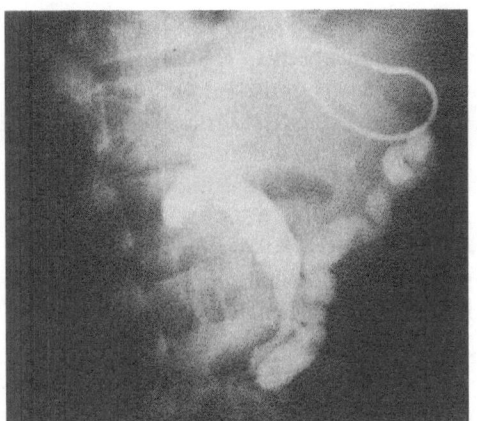

Fig. 16-6. Case 5. Periampullary carcinoma Both ducts are shown by tube cholangiography and pancreatography to be dilated and completely obstructed by tumor. (Fig. by permission of Radiology).

Aspiration biopsy of a pancreatic tumor is a useful method for the diagnosis of pancreatic cancer. Its sensitivity ranges from 81% to 86%. The reason for false negatives is aspiration from the fibrosis that accompanies carcinoma; even though the needle goes into the tumor, aspirated material is not adequate for cytology. Because of this, X-rays of the pancreatic duct become extremely important for differential diagnosis. Especially in patients in whom aspiration biopsy is negative but cancer is strongly suspected, or in whom differential diagnosis between cancer and chronic pancreatits with inflammatory mass formation is difficult, percutaneous pancreatography is very useful, as in *Case 4*. Moreover, when combined with other X-ray studies such as PTBD and hypotonic duodenography, the exact extent of the carcinomatous invasion can be determined, and the surgical method can be decided preoperatively. In patients in whom ERP is difficult due to previous gastrectomy, pancreatic divism or carcinoma of the papilla of Vater, percutaneous pancreatography is essential. There were no major complications, such as infection, bleeding, fistula or pancreatic necrosis. Only transient abdominal pain was seen in 2 cases (4.9%). A small collection may have resulted from a minor leak of pancreatic fluid due to the puncture. At first, cytological examination was thought to provide more information about the extent of carcinomatous invasion along the distal pancreatic duct, but its positivity rate was only 10% in patients with carcinoma of the head of the pancreas. Destruction of the intraductal malignant cells by digestive enzymes may be one of the causes of the low detectability. The concentration of amylase in the pancreatic fluid ranged widely. These values may help in evaluating the exocrine function of the non-cancerous remaining portion of the pancreas, therefore surgery should be carefully performed in order to avoid leakage at the pancreato-jejunostomy, especially in patients with a high concentration of amylase.

Ultrasonically guided percutaneous pancreatography is easy and safe, causing no major complications. Indications for this examination are: 1) patients in whom aspiration biopsy of the pancreatic mass is negative but cancer is strongly suspected, especially for differential diagnosis between cancer and chronic pancreatitis with inflammatory mass formation; 2) patients in whom ERP has failed or is impossible; and 3) patients in whom the exact extent of the cancer should be defined preoperatively with the additional aid of cholangiography or/and hypotonic duodenography.

References

Cooperberg P L, Cohen M M, Graham M. Ultrasonographically guided percutaneous pancreatography: report of two cases, *AJR* 1979; 132: 662.

Makuuchi M, Bandai Y, Ito T, Wada T. Ultrasonically guided percutaneous transhepatic cholangiography and percutaneous pancreatography. *Radiology* 1980; 134: 767.

Ohto M, Saotome N, Saisho H, et al. Real-time sonography of the pancreatic duct: application to percutaneous pancreatic ductography. *AJR* 1980; 134: 647.

Hancke S, Jacobsen G K. Puncture of pancreatic mass lesions. In: Holm H H, Kristensen J K, eds. Ultrasonically guided puncture technique. Copenhagen: Munksgaard and Baltimore: University Park Press, 1980: 61.

Ryu M, Uematsu S, Watanabe Y, et al. The meaning of ultrasonically guided puncture for the diagnosis of the pancreatic disease. *Jap J Gastroenterol Surg* 1981; 14: 45.

Puncture of gynecological masses

Flemming Jensen

Although ultrasonography plays a prominent role in the evaluation of patients with possible gynecological tumors, there are several difficulties in providing a specific diagnosis with ultrasonic imaging alone.

It may be rather easy to distinguish between normality and abnormality and also to judge the extension of disease, for example in staging of cervical uterine carcinoma. But many tumors in the pelvic region will present problems regarding their origin and true nature. Completely solid masses may not look very different, whether they are of *uterine, ovarian or intestinal origin* and although dynamic scanning simultaneous with vaginal palpation may solve problems of fixation, many patients still undergo surgery without a precise diagnosis.

Most solid tumors of the uterus are benign fibromyomas (Fig. 17-1), while almost all solid tumors of the ovary are malignant. With mixed cystic-solid lesions, the risk of malignancy increases with complexity, that is, the more prominent the solid components are (Fig. 17-2).

An exception to this rule of thumb is the ovarian *dermoid cyst,* which is very complex indeed, but benign.

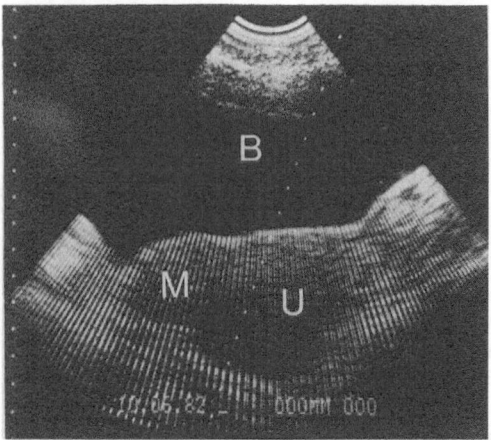

Fig. 17-1. Uterine myomas
The rounded, solid sound-attenuating tumors in proximity to the myometrium most often represent myomas, although their appearance may be more irregular than shown here (e.g. with calcifications or central hematomas). Curettage or fine needle biopsy can further render this diagnosis probable. B:bladder, U:uterus, M:myoma.

Purely or almost purely cystic tumors, especially the unilocular in very young females, are nearly always benign. They are either *neoplastic* as cystadenomas or *non-neoplastic* as follicular or luteal retention cysts.

Ultrasonically guided transabdominal fine needle biopsy can be useful if malignancy is thereby proven and the tumor extensive and clearly inoperable. But it is not much used when minor tumors

113

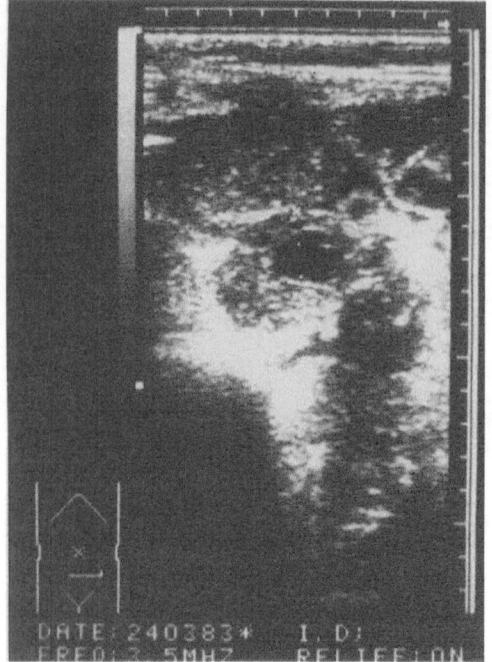

Fig. 17-2. Ovarian carcinoma
In this case the image is dominated by solid components, but the small cystic areas in this extensive lesion point to an ovarian origin.

than 10%. Maybe the confidence level could be raised by a more precise needle guidance.

Although percutaneous puncture anywhere else in the body carries an extremely low risk of tumor seeding (chapter 26) until now there seems to be a reluctant attitude to the procedure among gynecologists. They have feared the false negatives and, moreover, felt that the needle intervention in early cancer stages could itself alter the substaging (according to the FIGO-convention) because of the risk of leakage of tumor cells into the peritoneal cavity.

Thus until now, the role of guided biopsy has, not been in the *primary diagnosis* of gynecological tumors, but in patients already operated upon or otherwise treated for malignancy, and where *remanence*, *recurrence* or *extension* is to be proven before possible initiation of are seen, as these are potentially operable and the patient will undergo surgery anyway. This praxis without primary biopsy, tends to diminish the risk of false negative diagnoses. The cost is that many women are operated upon, when expectation or percutaneous emptying of a cyst could – retrospectively – have been chosen. Investigations covering biochemical analysis of cyst fluid to differentiate safely between *retention cysts* and *neoplastic cysts* are at present carried out and might in the future give the clue to an easier treatment of patients with cysts.

Based upon older literature, when puncture was performed without ultrasonic image guidance, there seems to be a not uncommon number of *false negative* cytological diagnoses; i.e., more

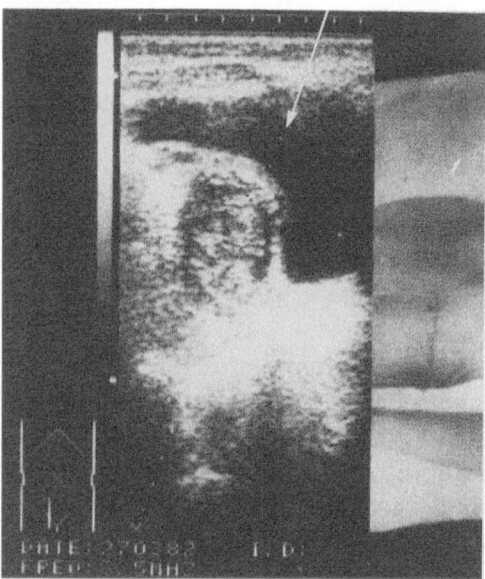

Fig. 17-3. Recurrent uterine carcinoma
The diagnosis was secured by a transvesical fine needle biopsy. The needle path corresponded to the dotted marker line, as a linear array biopsy transducer with oblique steering canal was used.

chemotherapy or irradiation (Fig. 17-3). Thus a number of so-called "second look" operations may be omitted (Fig. 17-4).

Fine needle biopsies are carried out transabdominally, possibly through the bladder (Figs. 17-5 & 17-6). The possibility of transvaginal puncture monitored with ultrasound also exists. In case of complex ovarian cysts with solid components, fine needle aspiration biopsy should be obtained from the solid components as well as from the cyst fluid.

Very small amounts of *ascites* are detectable and may be punctured for diag-

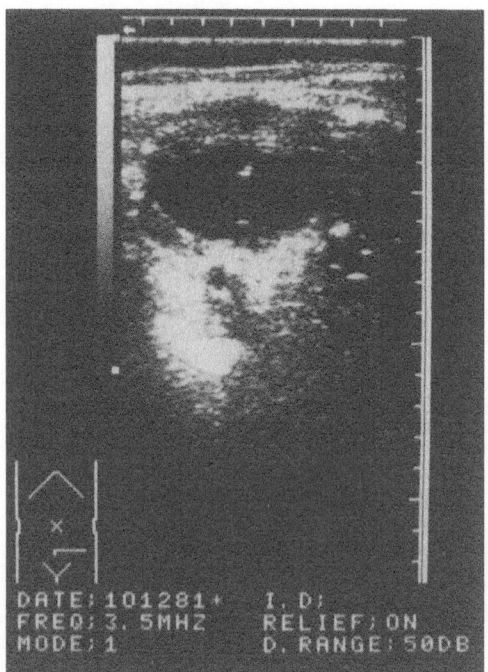

Fig. 17-5. Fine needle biopsy of cystadenocarcinoma
The needle-tip-echo is seen inside one cystic compartment of a recurrent ovarian cystadenocarcinoma.

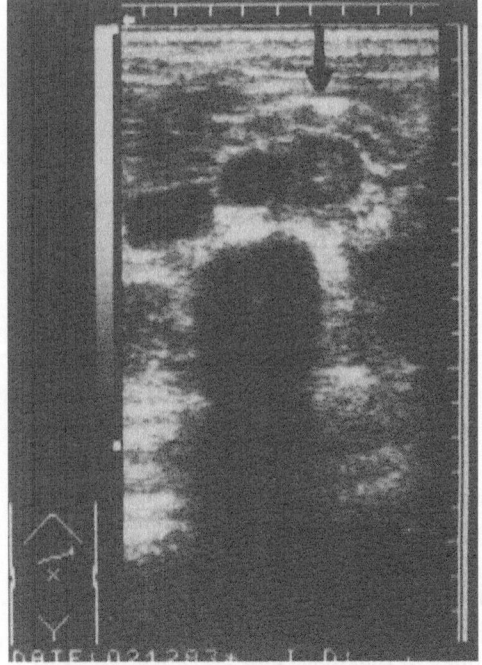

Fig. 17-4. Small para-aortic lymph node metastasis from cervical carcinoma
In this patient clinical examination, including vaginal palpation, did not reveal remaining disease, but by biopsy from this non-palpable, ultrasonically disclosed lesion, pathology was proven and the patient was given furhter therapy and the usual surgical proof ("second look" operation) was omitted.

nostic purposes, while the visualization of peritoneal carcinosis *per se* is less reliable. Sometimes, however, thickened intestinal loops and omentum are seen and aspiration biopsy from such areas is more likely to disclose the malignancy than cytological analysis of the fluid sediment (Fig. 17-7).

Ultrasonically guided puncture as a *therapeutic measure* in gynecology can, in certain cases, be performed for emptying of abscesses, postoperative hematomas or other fluid collections. It may also be used in connection with *percutaneous nephrostomy*, for example during radiation therapy or during chemotherapy, when ureteral obstruction is present.

115

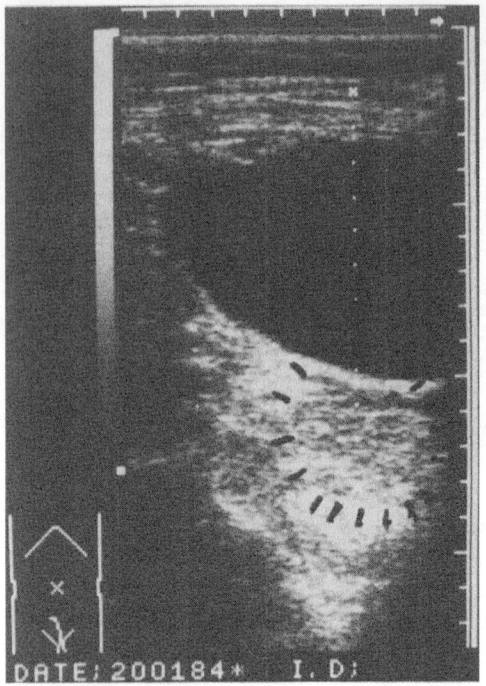

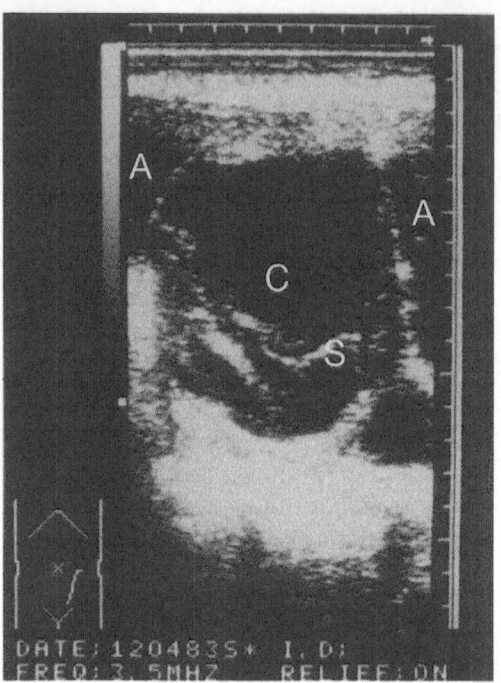

Fig. 17-6. Puncture route
This recurrent uterine sarcoma was verified by transvesical biopsy, but in cases with caudal site of the tumor (near the vaginal top) a transvaginal biopsy may as well be carried out.

Fig. 17-7. Large multicystic ovarian cancer with ascites
The cysts are separated by septa from which fine needle biopsy should be obtained. A:ascites, C:cyst, S:septum.

Reference

Berkowitz R S, Leavitt T, Knapp R C. Ultrasound-directed percutaneous aspiration biopsy of periaortic lymph nodes in recurrence of cervical carcinoma. *Am J Obstet Gynecol* 1978; 131: 906.

Gammelgaard J, Pedersen P H, Henriksen O B. Puncture of retroperitoneal, gastrointestinal and gynecological mass lesions. In: Holm H H, Kristensen J K, eds. Ultrasonically guided puncture technique, Chapter 15 Copenhagen: Munksgaard and Baltimore: University Park Press, 1980.

Kjellgren O, Ångström T, Bergman F, Wiklund D E. Fine-needle aspiration biopsy in diagnosis and classification of ovarian carcinoma. *Cancer* 1971; 28: 967.

Meire H B, Farrant P, Guha T. Distinction of benign from malignant ovarian cysts by ultrasound. *Br J Obstet Gynecol* 1978; 85: 893.

CHAPTER 18

Ultrasonically guided aspiration of human oocytes

Suzan Lenz

Ultrasound examination of the female reproductive organs is performed through a full urinary bladder in order to displace the air-containing intestines. The ovaries are seen beside, above or behind the uterus, and the cyclic changes are easily followed by repeated ultrasound examinations (Lenz 1984). Being fluid-filled, the follicles are depicted as echo-free spheric structures inside the echo-poor ovaries. The mature follicle measures 18–25 mm in diameter and is therefore an obvious target for puncture and aspiration, when the oocyte is to be collected and fertilized in the laboratory. This technique, called *in vitro* fertilization (IVF), is used in the treatment of infertility caused by severely damaged or absent Fallopian tubes or by oligospermia.

Pretreatment

To increase the woman's chances of conception, the follicular growth is stimulated in the treatment cycle to obtain several oocytes for fertilization and culture in the laboratory. In a normal cycle 1 or perhaps 2 oocytes will mature, but this number is insufficient for IVF considering the loss of oocytes and embryos at all steps in the procedure. The stimulation should start early in the cycle, when 5 to 10 follicles are recruited. At this time FSH will decrease in the normal cycle, and most of the selected follicles will become atretic. In a cycle naturally programmed to last 28 days, FSH will begin to decrease around the 4th day of the cycle, in a longer cycle later, and in a shorter one earlier. Therefore the start of the stimulation has to be calculated individually in each patient. Both clomiphene and hMG are used for stimulation in various combinations. No scheme seems to be superior to any other judged from the pregnancy rate obtained. A widely used practice is to begin with clomiphene 100 mg daily (50–200 mg) for 5 days and to continue with hMG 150 IU daily (75–300 IU) for 3 days. The dose and duration of treatment will depend on the individual response.

The follicular development can be followed by repeated estradiol measurement and ultrasound examination. The number and size of follicles are recorded by sonography and correlated to the estradiol level to determine the time of ovulation. It is important to remember that each woman will have her own pattern regarding maximal follicular size and growth rate, and that each stimulation therapy will influence this pattern in a different way. Furthermore, sound follicles will produce a great amount of estradiol and atretic or less sound follicles will produce smaller amounts. A sound follicle contains an oocyte, which can be fertilized. The quality of the follicle cannot, however, be predicted by its diameter. A possible positive sign is the appearance of an extensive, echo-poor cloud inside the follicle (Fig. 18–1) (Lenz et al. 1983). As only oocytes from mature, sound follicles can be fertilized this is a crucial point, and many prefer to await the natural LH-surge and then to aspirate the oocytes 36 hours after it starts. Others give hCG at a calculated time and aspirate the oocytes 36 hours later. The latter method implies the advantage that the aspiration can be performed at a chosen time of the day.

Equipment

A sector scanner is necessary for the aspiration and suitable for the monitoring. The monitoring can be performed by a compound scanner. A linear scanner will not reach the side walls of the pelvis. The sound beam should be focused at a depth of 9 to 12 cm, and the puncture canal should be long to provide firm, accurate guidance. One should be able to aspirate follicles as small as 14 mm in diameter.

A stainless stell needle is used for the puncture. The needle should have a lumen, which will allow the preovulatory expanded cumulus-mass to pass. The cumulus-mass is fortunately plastic and will pass a needle as thin as 19 gauge, but the recovery rate of oocytes per aspirated follicle will increase when a thicker needle is used. The needle has also to

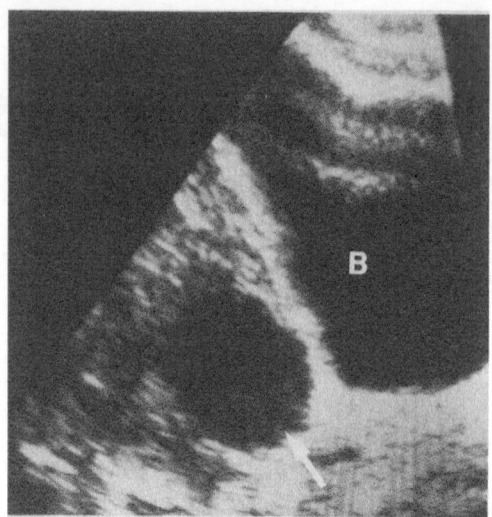

Fig. 18-1. Oocyte-containing follicle
Longitudinal echogram of a follicle containing an echo-cloud in the right, lower third (arrow). This cloud is presumed to indicate a sound oocyte. B: bladder. Centimeter scale is shown to the right.

be more rigid. A 16 gauge needle (outer diameter 1.4 mm, inner 1.0 mm) with stylet will fit. The needle point should be bevelled.

The follicular fluids are aspirated in plastic syringes specially manufactured for tissue culture.

Both needles and syringes have to be washed before use in extremely purified water freed of all possible organic and non-organic materials. The oocyte is sensitive to traces of many components, which may kill it. This has also to be considered when choosing the best method for sterilization of equipment. No gases should be used, only dry heat or gamma- and electronirradiation. The puncture transducer is sterilized by the same means.

Aspiration procedure and techniques

The oocyte aspiration is performed as

an outpatient procedure. The patient arrives with a full urinary bladder 1 hour before the scheduled time and is premedicated. A woman of average weight will be given 100 mg pethidin i.m. and 10 mg diazepam p.o. The size of the bladder is checked by ultrasound before the premedication takes effect. If the bladder is too distended the patient is asked to void an estimated volume. This does not seem to be a problem for these patients who have practised bladder filling and partial voiding during the monitoring of the cycle. The puncture route is then chosen to avoid scars on the abdomen and also, by means of the sonograms, the bowel and thick adhesions in the abdomen. The puncture should be as perpendicular to the skin as possible to avoid deviation of the needle. The path should go through the bladder and not be tangent to the bladder wall, because this is painful and might provoke bladder bleeding. The puncture can, however, be performed without traversing the bladder, if the ovary is fastened by adhesions to the abdominal wall. The local anesthetic is given according to the chosen puncture route and applied to all layers of the abdominal wall and also to the front wall of the bladder. Puncture of the follicles should be avoided and the needle should be withdrawn as soon as urine is aspirated.

The skin is sterilized by means of 60–70% alcohol. The skin detergent might come in contact with the follicular fluid, and iodine or chlorhexidine could damage the oocytes. The gloves are rinsed in purified water, and the puncture transducer is placed on the abdomen after draping. Skin contact is provided by sterile oil or, better still, by purified water.

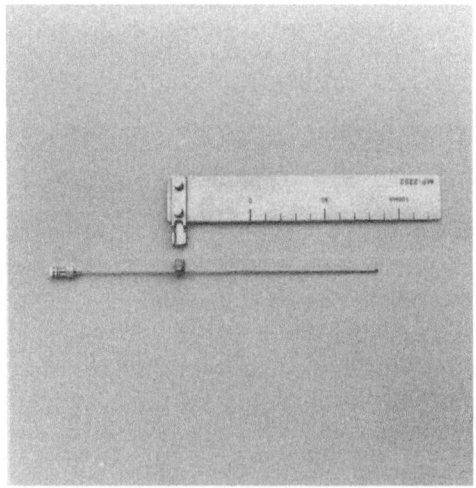

Fig. 18-2. Marking of puncture depth
The needle with stopper shown beside the puncture ruler. The stopper is placed in the groove and the needle is pushed until the needle tip is adjusted at the desired measurement. Then the stopper is fixed to the needle. The ruler has included the depth of the puncture transducer.

The first follicle to be punctured is pointed out with the guide line, and the distance from the skin to the bottom of the follicle is measured. This distance is by means of a ruler marked on the needle with a stopper (Fig. 18-2). If the needle point echo should disappear on the screen in the echo-poor area of the ovary or its surroundings, this stopper will help to hit the correct depth.

The needle is passed into the bladder lumen, where the puncture direction is easily controlled. The patient should be warned that her full coorporation in lying still is essential. In case the follicle has come out of focus, the needle should be withdrawn completely and a new puncture performed. Before penetrating the ovary it is advisable to tell the patient that she may feel a little pain, and that it is very important she does not move. The needle is passed into the fol-

119

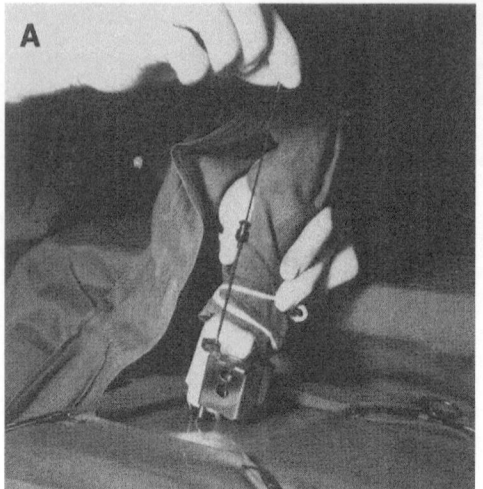

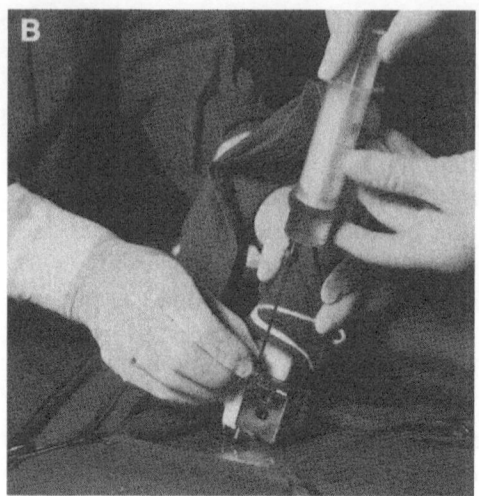

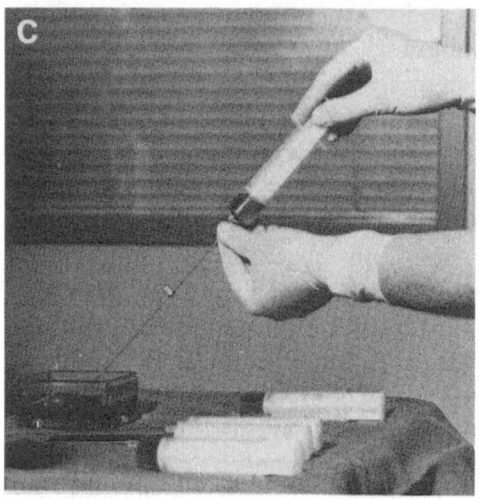

Fig. 18-3. Aspiration procedure
Steps in the aspiration procedure:
A: The stylet is removed when the needle tip is inside the follicle – note the stopper at the needle.
B: Immediately after an empty syringe is attached to the needle and the follicular content is aspirated.
C: After aspiration and flushing the needle and syringe are removed and a needle flush is carried out. See text for further explanation.

licle with a rapid movement, to minimize pain and prevent the ovary from moving away. The stylet is retracted (Fig. 18-3A) and an empty syringe attached to the needle. This has to be done very quickly. Once the system is closed there is plenty of time to aspirate the follicular content gently (Fig. 18-3B). The collapse of the follicle is observed on the screen. When the follicle is empty, the syringe is replaced by another syringe containing warm (37°C), heparinized culture medium. The aspirate is then replaced by exactly the same volume of medium injected into the follicle. The expansion of the follicle can be monitored on the screen. If the fluid is leaking due to a wrong position of the needle tip or to overloading, the injection should be stopped. The patient will experience pain, and the fluid is lost. The fluid is reaspirated and the collapse of the follicle is followed on the screen. Flushing can be repeated many times. The needle and syringe are withdrawn and the procedure is completed with a needle flush (Fig. 18-3C). Repeat the procedure for each follicle.

The aspirates and flushes should be delivered as fast as possible to the laboratory for oocyte detection and incubation. The fluids should not be exposed to cooling, bright light or normal atmosphere for too long a period as the oocytes are extremely sensitive. An oocyte should be found in at least 60% of the follicles and with the stimulation described this means that oocytes will be obtained in 90–95% of the patients.

As soon as the punctures are completed the patient is asked to void. Some patients may have difficulty in emptying their bladder due to pain, overdisten-sion and pethidin, but it is seldom necessary to use a catheter. In about half of the cases there will be small blood clots in the urine, and in 5% of the cases there will be hematuria in the first couple of voidings. In all cases it seems to disappear after a few hours. The patient can leave the department after 3 to 4 hours of observation. No other complications to the procedure are known.

References

Lenz S. Ultrasonic study of follicular maturation, ovulation and development of corpus luteum during normal menstrual cycles. *Acta Obstet Gynecol Scand* 1985; 64: 15.

Lenz S, Lauritsen J G, Lindenberg S. Significance of the ultrasonic morphology of preovulatory ovarian follicles prior to in vitro fertilization. *Acta E Fert* 1983; 14: 305.

Lenz S, Bang J, Lauritsen J G, Lindenberg S. Ultrasonically guided aspiration of oocytes for in vitro fertilization using a plain needle and syringe under local anesthesia. *Infertility* (in press).

Intrauterine needle diagnosis

Jens Bang

Since Bevis demonstrated the value of examining the amniotic fluid for assessment of erythroblastosis in the 3rd trimester, indications for amniocentesis have been greatly extended. Ultrasound scanning plays a central role in enhancing the effectiveness and the safety of this diagnostic technique and makes it possible to supplement the simple amniocentesis with more complicated diagnostic procedures such as fetal blood sampling and biopsy taking from the fetus.

Indications

In the 2nd trimester, amniocentesis is now widely used for genetic diagnosis. Chromosomal disorders may be diagnosed by cytogenetic examination of fetal cells from the amniotic fluid. Inborn errors of metabolism may be diagnosed by studies of fetal cells or, in a few cases, by examination of amniotic fluid. Neural tube defects may be revealed by the study of the alpha-fetoprotein content of the amniotic fluid and examination of acetylcholinesterase and concanavalin A. Fetal sex may be determined in families with X-linked diseases. The number of 2nd trimester amniocenteses is growing rapidly. The number of pre- natal genetic diagnoses carried out has been limited by lack of funds in most countries. In Denmark, however, a special appropriation was given in October 1978 to cover genetic caunselling and diagnostic amniocentesis offered to pregnant women in the following situations: 1) age 35 years or more; 2) previous birth of a child with a chromosome anomaly; 3) when a pregnant woman or the father has a translocation or another chromosomal aberration; 4) previous birth of a child with a metabolic disease or an X-linked disease; 5) if elevated alpha-fetoprotein is found in serum; or 6) previous birth of a child with myelomeningocele or anencephalia or if neural tube defects exist in the family. Second trimester amniocentesis is made for these purposes in 5–6,000 cases per year in Denmark. (Population 5 million).

In some diseases fetal tissue is required to identify specific genetic characteristics by biochemical or cytogenetic analysis of amniotic fluid. Fetal blood can be used for prenatal diagnosis of hemoglobinopathies, coagulation, metabolic and cytogenetic disorders, immunodeficiencies or fetal infections. Fetal skin biopsy is performed for the diagnosis of genodermatoses such as epidermolysis bullosa letalis or epidermyolysis bullosa dystrophica. It is also possible to evaluate liver-specific enzyme deficiencies by means of biopsies from the fetal liver. Such procedures can be guided by fetoscopy, but can also be performed as percutaneous needle biopsies under ultrasonic guidance.

The methods used until now cannot give reliable information until the 16th to 20th week of pregnancy. This delay causes considerable anxiety, and in cases where the fetus is found to be abnormal, this subjects the mother to the increased risks, both physical and emotional, of a 2nd trimester abortion. To eliminate some of these problems, recent interest has been focused on developing a method of sampling fetal tissues in the 1st trimester. It is possible to sample chorionic tissue by needle biopsy transabdominally or by means of a catheter introduced through the cervical canal. Both procedures are done under ultrasound guidance. The indications in the 1st trimester are fetal sex determination, karyotyping, enzyme analysis and analysis by means of recombinant DNA technology.

Third trimester amniocentesis is discussed in chapter 20.

Method

Amniocentesis should be carried out under ultrasound guidance because this increases the success rate. In a Canadian collaborative study by Simpson et al., the success rate of the first attempt improved from 76% to 86% after placental location by ultrasound, and Crandon & Peel achieved an increase in the success rate of amniocentesis from 80% to 99.6% after the introduction of ultra-

sound guidance. The incidence of blood contamination in the amniotic fluid was similarly reduced from 43.5% to 17.6%.

Since 1970, we have used an ultrasound puncture transducer allowing amniocentesis under simultaneous visual ultrasound guidance. Until 1980 we used bi-stable equipment with a special puncture transducer (Holm et al.). Since 1980 we have used real-time equipment, Aloka SD 256 mounted with a puncture transducer.

The transducer is sterilized in hibitane solution and the skin is swabbed with iodine. Local anesthesia is never used (most patients find that the puncture hurts less than the taking of a blood sample). Ultrasonic scanning is performed to ensure: 1) presence of an intrauterine pregnancy with a living fetus; 2) estimation of the gestational age by BPD measuring; 3) presence of single gestation or twins (and diagnosis of mono- or diamniotic gestation); 4) abnormalities of the fetus; 5) the amount of amniotic fluid; 6) location of the placenta.

A marker line on the oscilloscope indicates the needle path which should intersect a suitable collection of amniotic fluid. The procedure should be carried out under simultaneous ultrasonic monitoring since the fetus may move and the uterus may contract or rotate.

The needle used is 1.0 mm/150 mm. To ensure penetration of the membranes the needle is inserted to the appropriate depth in a single rapid movement. Routinely, 2 ml of amniotic fluid are aspirated for analysis of alfa-fetoprotein and, in another syringe, 15 ml for karyotyping. In the case of twins, it is normally possible to puncture on each side of the membrane between the 2 gestational sacs. If it is difficult and uncertain, a special technique with injection of 2 ml Congo red 1.5% just after the aspiration is used. Puncture of the 2nd gestational sacs is then attempted. If the amniotic fluid removed is of normal colour, samples from both cavities have been obtained. It must be mentioned that the frequency of spontaneous abortion is higher in twin pregnancies than in single pregnancies, and as a consequence of this, all patients in whom twins are discovered receive secondary genetic counselling before amniocentesis is decided upon. They have to be informed about the higher risk of amniocentesis and its possible result: 2 normal, 2 sick, or 1 normal and 1 sick fetus. The counselling involves the discussion of possible selected feticide.

When fetal blood sampling is indicated, a puncture transducer with variable angularity of the needle path is used (Fig. 19-1). The skin is swabbed twice with iodine and the region is covered with sterile drapes. The position of the transducer is changed until the marker line on the oscilloscope intersects the left ventricle of the fetal heart. A guide needle (1.2 × 150 mm) is introduced into the fetal thorax and through this a fine needle (0.6 × 180 mm) is inserted into the lumen of the left ventricle. Then 0.5–1 ml of fetal blood is withdrawn.

This procedure is used in the 18th–20th week of pregnancy. In the 3rd trimester the same technique has been used for introduction of the fine needle into the hepatic part of the umbilical vein with the aim of fetal intravenous transfusion or blood sampling.

In case fetal skin biopsy has been indicated, this has been performed, guided ultrasonically. A Tru-cut® needle

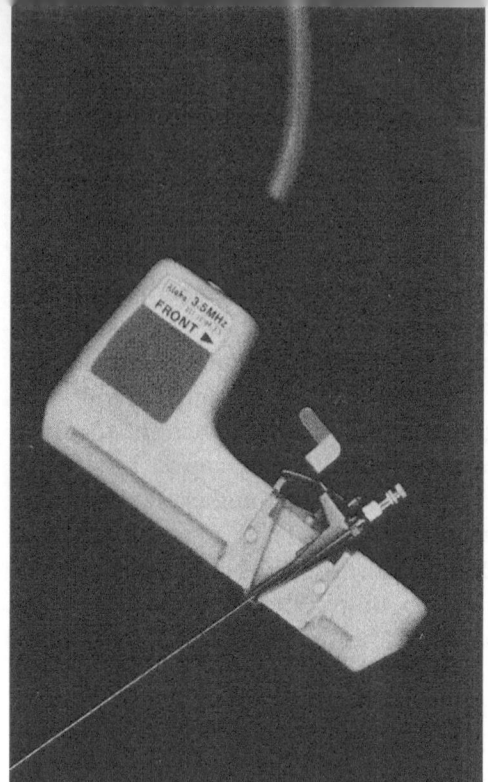

Fig. 19-1. Puncture transducer
The needle is placed in the steering device that allows different angularity.

is introduced through the puncture transducer into the amniotic cavity. The needle is opened and guided to the skin surface. When the needle is closed, while pressed against the skin, a biopsy from the fetal skin is obtained (Fig. 19-2).

Results

In the period from March 1973 to September 1982 we have performed 9,785 ultrasonically guided amniocenteses. Out of these, 176 (1.8%) were macroscopically blood stained. In 127 cases (1.3%), 2 punctures were made on the same day, and in 31 (0.3%) a 2nd amniocentesis was made a week later.

These figures demonstrate that the method is suitable for sampling of pure

Table 19-1. Outcome of 1177 pregnancies after amniocentesis

	Born alive		Stillborn	Spontaneous abortions	Induced abortions	Total infants[x]
	> 2500 g	< 2500 g				
No. (%) of infants	1039 (87.8)	79 (6.7)	11 (0.9)	28 (2.4)	26 (2.2)	1183 (100.0)

[x]Six cases were twin pregnancies.

amniotic fluid providing the basis for an accurate and fast analysis. They also demonstrate that the need for more than 1 puncture is very slight. A collaborative study from Canada (Simpson et al.) showed this to be important as regards the frequency of abortions. However, it is necessary to evaluate the safety of amniocentesis by estimation of its possible influence on the frequency of spontaneous abortions, premature deliveries and the rate of malformations.

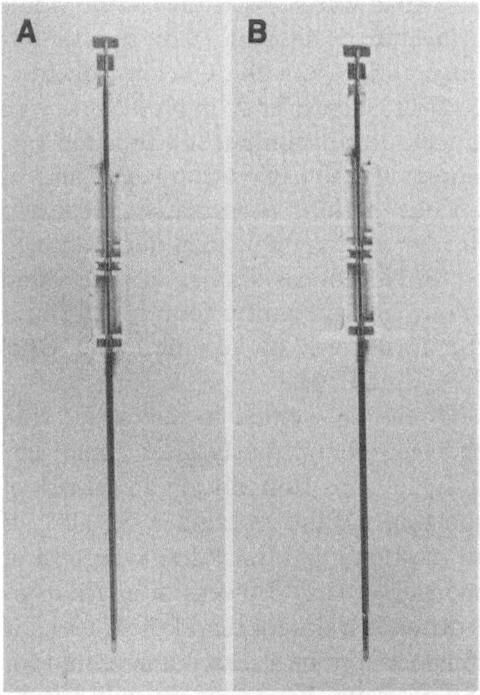

In a report from the English Medical Research Council, an increased risk of 1–1.5% for spontaneous abortion and an increased risk for congenital dislocation of the hip was found. However, these results are compiled from a multicenter study, and a large number of the samplings have been performed without ultrasonic facilities. It is difficult to evaluate the material and impossible to compare it with ultrasonically guided amniocentesis.

Table 19-1 demonstrates the outcome of the first 1177 pregnancies in which second trimester amniocentesis was performed. All cases have been followed after delivery. Of 1183 infants, 1118 (94.5%) were born alive: 1039 (87.8%) weighed over 2500 g, and 79 (6.7%) under 2500 g. Twenty-six terminations were performed, 13 because of chromosomal abnormalities, the remainder because of raised alfa-fetoprotein concentration, rubella infection, male fetuses in families with X-linked diseases, maternal diabetes, and adrenogenital syndrome. In 1 of the 4 cases of twins not detected by ultrasound before amniocentesis, a liveborn infant with Down's syndrome was delivered, and in 1 case where the cells did not grow, a mongoloid child was born. Otherwise no visible malformations caused by chromosome abnormalities were found. No damage caused by the amniocentesis was recorded after delivery.

Fig. 19-2. Tru-Cut® needle for skin biopsy
A: The needle closed. B: The open needle showing the slit which can be pressed against the surface of the skin.

125

Table 19-2. Indications for amniocentesis and incidence of spontaneous abortion

	Maternal age >40	Maternal age 35–39	Other (maternal age <35)	Total
No. of amniocenteses	256	469	454	1179[x]
No. (%) of spontaneous abortions	8 (3.1)	14 (3.0)	6 (1.3)	28 (2.4)

[x]Includes 2 cases in which twins were diagnosed before amniocentesis.

Twenty-eight pregnancies (2.4%) ended in spontaneous abortion (Table 19-2). Amniocentesis was usually performed during or after the 16th week of gestation, but in about one quarter of the cases it was performed earlier. The earlier the amniocentesis is performed, the more likely the mother is to have spontaneous abortion, independent of the amniocentesis. In 3 of the 28 cases (10.7%) repeated punctures were necessary owing to difficulties in obtaining fluid or in culturing cells. The overall incidence of repeated punctures in the series was 3%. Only 6 of the spontaneous abortions occurred in women under 35 years of age (Table 19-2). The mean age of all 28 cases was 35 years, range 23-49.

Three of the 1177 women (0.3%) had their first symptom of abortion within 8 days after amniocentesis. Eight (0.7%) had their first symptom between 8 days and 3 weeks and 17 (1.4%) after 3 weeks. Eleven of the 28 women already had conditions associated with an increased risk of abortion. A further 8 probably had such conditions. Risk factors noted in the series were bleeding, cervical insufficiency, hydrocephalus, twins, intrauterine device *in situ,* and abnormalities of the placenta. None of the spontaneously aborted fetuses had a chromosomal abnormality. Simpson et al. have shown that the risk of abortion increases when amniocentesis proves

difficult to perform or needs to be repeated.

From March 1973 to September 1980 we performed chromosome analysis on 5372 amniotic fluid samples. In 120 cases (2.22%) spontaneous abortions occurred.

The incidence of fetal loss after amniocentesis may be assessed indirectly by comparison with figures for prematurity and abortion.

In our series the incidence of premature birth was 6.7% while for the whole of Denmark during 1976 it was 6% (Danmarks Statistik). Obel in Copenhagen and Shapiro et al. in New York, who analyzed the spontaneous abortion rate among women of various ages and of various periods of gestation, reported incidences among women not examined by amniocentesis similar to that found in our series. Shapiro found 3.2% after the 15th week of gestation and Obel 2.7% in all ages.

Of the 28 women in our series who aborted spontaneously after amniocentesis, 11 were probably already at risk of aborting. Of the remainder, 8 (0.7% of all mothers) had their first symptom of abortion within 3 weeks after the procedure. If a time interval of 3 weeks is considered to be a reasonable limit, these 8 cases may have been causally associated with amniocentesis. Three women without risk factors associated with abortion had the first symptom within

1 week after the procedure. If this is regarded as a reasonable time limit, 0.25% of all mothers may have aborted as a result of amniocentesis.

Our findings suggest that amniocentesis in early pregnancy does entail a small risk of spontaneous abortion. Nevertheless, about 0.5% of newborn babies have chromosomal abnormalities. Confining amniocentesis to women over 35 results in only 20–30% of cases of Down's syndrome being detected before birth, as demonstrated by Mikkelsen et al.

The overall incidence of abortion after the 15th week of pregnancy and independent of maternal age is 2–2.5%. Our findings suggest that, if all pregnant women undergo amniocentesis, the incidence will increase by 0.3–0.7%. This, however, must be weighed against the 70–80% of cases of Down's syndrome and even higher proportions of other genetic disorders that go unrecognized before birth.

The main indication for fetal blood sampling in genetic diagnostics in Scandinavia is hemophilia, A or B. The usual technique for fetal blood sampling is by means of fetoscopy. By this method mixed samples of blood from the mother and the fetus are often obtained. It is acceptable in the diagnosis of hemoglobinopathia, but not for hemophilia. The number of cases where the operation is indicated are few in Denmark, and we found it difficult to achieve the necessary routine. After several years of experience with ultrasonically guided amniocentesis, we have developed a technique with ultrasonically guided fetal heart puncture for sampling of fetal blood.

Fetal heart puncture with blood sampling has been carried out in 40 patients admitted for termination of pregnancy in the 16th–18th week. All the fetuses were carefully examined by a pediatric pathologist, especially concerning the thorax and heart. In a few cases there was a minimal amount of blood in the pericardiac cavity, but there were no other signs of lesion in the thorax or cardiac region. Until now, we have performed 35 punctures (Table 19-3). In the 24 cases where a normal fetus was found, 18 normal children have been delivered, 4 continued their pregnancy, but have not delivered yet and 1 fetus was aborted spontaneously 3 weeks after the procedure. The mother has had another pregnancy, and she aborted at the same gestational age. In 3 cases of twins where 1 of the fetuses was abnormal (2 cases with Down's syndrome and 1 with hydrocephalus) the technique with

Table 19-3.Diagnostic fetal blood sampling

Indications	Pregnancies total	No. of sick fetus	No. of normal fetus
Hemophilia A	22	7	15
Hemophilia B	3	2	1
Chronical			
Granulomatous disease	2	1	1
Rhesus immunization	6		6
Karyotyping	1		1
Rubella	1	1	
	35	11	24

puncture of the fetal heart has been used for injection of air into the abnormal fetus. One of the cases with Down's syndrome delivered a normal child by Caesarean section in the 37th week. The other 2 patients aborted spontaneously 3 and 4 weeks after the air injection.

In 3 cases we have performed skin biopsy. Two fetuses were found to be abnormal with epidermolysis bullosa, and 1 was suspected of this diagnosis but proved normal. The patient aborted 5 weeks after the procedure.

In the 3rd trimester in cases with rhesus iso-immunization, it is sometimes necessary to perform intrauterine, intraperitoneal transfusion. They are performed by means of ultrasonic guidance. When there is a severe rhesus hemolytic disease with hydrops including ascites, we have used an ultrasound guided fetal intravenous transfusion into the hepatic part of the umbilical vein. By means of this procedure we have suc-

ceded in giving 3 very severe cases transfusions with an interval of 1 week, 3 times in each pregnancy and delivering by Caesarean section in the 31st–33rd week of gestation.

Knowledge of various factors such as the optimal amount of transfused blood and its cardiovascular effect is important, as are the variables in fetal blood. We hope that direct intravenous fetal transfusion will improve the prognosis of high-risk, rhesus sensitized fetuses that cannot be saved by traditional methods.

In relation to the intrauterine transfusion, we have followed the patient by blood analysis weekly, for the estimation of hemoglobin and bilirubin. Furthermore, we analyze the ratio between fetal and adult erythrocytes by means of a Coulter Counter with a chanalyzer. The blood samples are obtained from the umbilical vein.

References

Bevis D C A. Composition of liqour amnii in haemolytic disease of newborn. *Lancet* 1950; 2: 443.

Simpson N E, Dallairi L, Miller J R, et al. Prenatal diagnosis of genetic disease in Canada: Report of a collaborative study. *Can Med Assoc J* 1976; 115: 739.

Crandon A J, Peel K R. Amniocentesis with and without ultrasound guidance. *Br J Obstet Gynecol* 1979: 86: 1.

Holm T H, Kristensen J K, Rasmussen S N, Nortehved A, Barlebo H. *Ultrasonics* 1972: 10: 83.

Bang J, Nortehved A. A new ultrasonic method for transabdominal amniocentesis. *Am J Obstet Gynecol* 1972; 114: 599.

Bang J, Nielsen H, Philip J. Prenatal karyotyping of twins by ultrasonically guided amniocentesis. *Am J Obstet Gynecol* 1975;

123: 695.

Philip J, Bang J. Outcome of pregnancy after amniocentesis for chromosome analysis. *Br Med J* 1978; 2: 1183.

Obel E. Risk of spontaneous abortion following legal abortion. *Acta Obstet Gynecol Scand* 1980; 58: 131.

Rodeck C H, Nicolaides K H. Ultrasound guided invasive procedures in obstetrics. *Clin Obstet Gynecol* 1983; 10: 515.

Shapiro S, Jones E W, Densen P M. A life of pregnancy terminations and correlates of fetal loss. *Milbank Memorial Found Quart* 1962; 40: 7.

Mikkelsen M, Fischer G, Stene J, Stene E, Petersen E. Incidence study of Down's syndrome in Copenhagen 1960–71, with chromosome investigation. *Ann Hum Genet* 1976; 40: 177.

CHAPTER 20

Amniocentesis, late pregnancy

Jan Fog Pedersen

Before ultrasound became available amniocentesis was performed blindly, guided by a palpatory impression of the position of the uterine content, but today amniocentesis should always be carried out under ultrasound guidance.

Indications

Amniocentesis in late pregnancy is performed in order to determine the contents in the amniotic fluid of surface active phospholipid, as an indicator of fetal pulmonary maturity, or, more rarely, to determine the bilirubin concentration in cases of fetal hemolytic disease.

Knowledge of fetal pulmonary maturity is essential when induced delivery or elective Cesarean section is considered, in order to avoid respiratory distress syndrome (RDS) of the newborn. Usually the lecithin/sphingomyelin (L/S) ratio is determined, and when this is above 2.0, the risk of RDS is very low. However, when the fetal age is 38 weeks or more, the risk of developing RDS is in any case extremely low (Frigoletto et al.). With the increasing use of 2nd-trimester ultrasound examination in cases of uncertain chronology, the number of pregnancies with questionable gestational age is diminishing, and the number of amniocenteses for L/S ratio determination is consequently diminishing. Even maternal diabetes mellitus, which has been a well-established indication for L/S ratio determination, is no longer an indication for amniocentesis in the 3rd trimester.

Thanks to prophylaxis, severe hemolytic disease (Rh-disease) is now a rare condition, and the observation and treatment of these patients, which may involve intrauterine transfusion (chapter 21), is concentrated in special centers.

As a result the number of late amniocenteses performed at most obstetrical departments is low and decreasing, in contrast to the number of 2nd trimester amniocenteses. Evaluation of the incidence of complications to 3rd trimester amniocentesis is therefore more difficult than is the case in the 2nd trimester. There are no series comparable to the reported collaborative studies of 2nd-trimester amniocentesis. Still, ultrasound guided 3rd trimester amniocenteses must be considered a very safe procedure and with experience in 2nd trimester amniocentesis it should be possible to adequately perform the occasional 3rd trimester amniocentesis.

Technique

The amniocentesis should be performed under ultrasound guidance, and it is probably wise to use the same technique as one is familiar with for 2nd trimester amniocentesis. The major difference is that there is relatively less amniotic fluid in 3rd trimester and the fetus is heavier, so that it is not as easily pushed away by the needle if inadvertently hit.

Many modifications of ultrasound guidance have been suggested, static B-scanning with or without a puncture transducer with a central canal, or dynamic scanning with or without mechanical guidance of the needle. Each technique has its advantages and drawbacks. The important thing is that the amniocentesis is performed using continuous ultrasound monitoring, so that the tip of the needle is kept in sight and it is ensured that the fetus has not moved into the intended path of the needle.

Whichever ultrasound guidance technique is chosen there are certain hints concerning amniocentesis that can be taken as general:

1. Initially a survey scanning is performed to locate a suitable pocket of amniotic fluid and define the optimum site of puncture. It is usually better to insert the needle close to the fetus, aiming away from it, rather than to puncture further away from the fetus but aim towards it. Preferably, the needle should not penetrate the placenta. If the only way to the amniotic fluid is through the placenta, it should be considered whether the benefit from the amniocentesis outweighs the risk of bleeding.

2. The area of sterilized skin should be rather large, to enable adjustment of the proper puncture site in case the fetus has moved.

3. A thin needle should be used to minimize the irritation to the uterus as well as any vascular damage. A disposable infant lumbar needle, 0.9 mm outer diameter (gauge 20), 90 mm long is well suited, and is in most cases long enough also to pass through a 3–4 cm device for puncture guidance. If no mechanical guidance is involved a shorter and thinner needle, for example 0.7 mm outer diameter (gauge 22), 70 mm long may be used.

4. The needle should be inserted into the amniotic cavity somewhat abruptly, in order to penetrate the membranes instead of just separating the membranes from the uterine wall, the so-called tenting phenomenon.

5. The thin needle causes very little pain, and no local anesthetic is needed.

Results and complications

Review of the literature shows a rate of failed amniocenteses of 5–10%, depending on the amount of amniotic fluid and the position of the fetus and the placenta (Galle & Meis).

The risk of spontaneous labor as a consequence of amniocentesis is related to the proximity to term, and increases from less than 1% in the middle of the 3rd trimester to 10% after 38 weeks. In approximate figures the reported rate of rupture of the membranes within 24 hours is 3%, of bloody taps 5% and of fetal injury 1% (Picker et al., Galle & Meis). Improvements in image quality and increased familiarity with ultrasound guided punctures will tend to reduce these figures.

Maternal complications are extremely rare, bleeding due to laceration of deep epigastric or uterine vessels being the most serious. The risk of this complication can be reduced by avoiding far lateral approaches whenever possible.

References

Bang J, Northeved A. A new ultrasonic method for transabdominal amniocentesis. *Am J Obstet Gynecol* 1972; 114: 599.

Benacerraf B R, Frigoletto F D. Amniocentesis under continuous ultrasound guidance: A series of 232 cases. *Obstet Gynecol* 1983; 62: 760.

Frigoletto F D, Phillippe M, Davies I J, Ryan K J. Avoiding iatrogenic prematurity with elective repeat Cesarean section without the routine use of amniocentesis. *Am J Obstet Gynecol* 1980; 137: 521.

Galle P C, Meis P J. Complications of amniocentesis. A review. *J Reprod Med* 1982; 27: 149.

Gross T L, Sokol R J, Wolfson R N, Kazzi N J. When is an amniocentesis for fetal maturity unnecessary in nondiabetic pregnancies at risk? *Am J Obstet Gynecol* 1984; 149: 311.

Jeanty P, Rodesch F, Romero R, Venus I, Hobbins J C. How to improve your amnioncentesis technique. *Am J Obstet Gynecol* 1983; 146: 593.

Pedersen J F. Percutaneous puncture guide by ultrasonic multitransducer scanning. *J Clin Ultrasound* 1977; 5: 175.

Picker R H, Smith D H, Saunders D M, Pennington J C. A review of 2,003 consecutive amnioncenteses performed under ultrasonic control in late pregnancy. *Aust N Z J Obstet Gynaecol* 1979; 19: 83.

Yeast J D, Garite T J, Dorchester W. The risk of amniocentesis in the management of premature rupture of the membranes. *Amer J Obstet Gynecol* 1984; 149: 505.

CHAPTER 21

Fetal therapy

Michael Manco-Johnson, William Clewell, Dolores Pretorius & David Manchester

The ultrasonic diagnosis of fetal congenital anomalies has now reached such a sophisticated level that many fetal diseases and malformations can be accurately diagnosed. Detection of fetal disease has dramatic effects on the management of the pregnancy, fetus and newborn. Harrison has divided fetal diseases into categories based on method of treatment; selective pregnancy termination, delivery at term with planned treatment of the neonate, premature delivery in order to institute treatment, Cesarean delivery in order to reduce trauma to the infant, and finally, fetal treatment. This last category includes conditions that cause progressive damage to the fetus, impair further normal development or threaten fetal life at a stage of pregnancy when premature delivery is not a reasonable option. Though this is a very small segment of the patient population, it has excited immense interest among scientific and lay communities. At times this interest has been out of proportion to the true importance of these developments. Fetal therapy is not a new concept. Long before the first attempts at direct fetal treatment, obstetricians attempted to improve fetal well-being by improving maternal health. In 1963 Liley performed the first intrauterine fetal transfusion for erythroblastosis fetalis. This revolutionized the treatment of this disorder. Intrauterine transfusion remained the only direct treatment for fetal disease until recently and serves as a useful model for more recent attempts at fetal therapy.

Prerequisites for fetal therapy

Accurate diagnosis

The most important prerequisite for fetal treatment is an accurate diagnosis. Not only must congenital anomalies be accurately identified, but additional and complicating malformations or diseases must be identified or ruled out. To this end, family histories and consideration of genetic amniocentesis should always be included. Much of what can be learned about a fetus, however, will depend upon the skill and experience of the ultrasonographer and a thorough and complete examination of both the fetus and the mother. This includes visualization and assessment of the fetal brain, heart, pleural cavity, liver, kidneys, bladder, GI tract, skin, spine and long bones. No organ system should be omitted.

Lesion responsive to therapy

Any lesion being considered for *in utero* treatment must be known to respond to the proposed therapy and, if alleviated, allow more normal fetal development to proceed. The presence of additional diseases or malformations should probably be considered a contraindication to fetal treatment. Additional malformations may render the proposed treatment futile. Applying experimental treatments to the fetus with multiple problems will confuse the interpretation of results and should be avoided if possible. Since premature delivery with neonatal treatment is an acceptable, if not preferable, approach to many of the diseases considered for *in utero* treatment, it must be carefully considered in every case.

Lesion is progressive

It should be demonstrated with as much certainty as possible that the lesion is progressive. One cannot justify experimental treatment if the disease appears to be stable. Treatment should be attempted only if there is a clear danger of continued and irreversible damage to the fetus.

Capable medical team

A multi-disciplinary team should be assembled which should include an obstetrician experienced in fetal diagnosis and intrauterine transfusion techniques, an ultrasonographer experienced in the diagnosis of fetal anomalies and percutaneous surgical techniques, a pediatric surgeon experienced in the neonatal management of the disease in question and a neonatologist who will manage the infant after delivery. The input of experienced dysmorphologists and clinical geneticists is also invaluable. These individuals should all concur in the proposed plan for treatment.

Concurrence of institutional review board

Since interventions involving the fetus are almost all experimental, an institutional review board (IRB) must be consulted and approve any program of fetal treatment. The ethical issues involved in fetal experimentation are complex and careful consideration must be given to fetal, maternal, and professional rights.

Investigative soundness

Prior to embarking on experimental treatment, provision must be made for follow-up and evaluation of results. Failure to make such provisions or failure to report results to the medical community constitutes a breach of good research practice and could be construed as unethical.

If these prerequisites are met, the following medical and surgical fetal problems can be considered for *in utero* therapy.

Erythroblastosis fetalis

With the development of Rh(D) immune globulin (RhoGam), the number of Rh-sensitized patients has greatly decreased, but erythroblastosis fetalis nevertheless remains a serious clinical problem, especially for the major referral centers. Intrauterine transfusion (IUT) was first developed by William

Liley in 1963 and the technique was significantly improved with fluoroscopic guidance. With the advent of high resolution real-time ultrasound scanners, intrauterine transfusion is now even safer and faster and the needles can be placed very accurately in the lower fetal abdomen, avoiding critical organs such as the fetal aorta, liver and spleen.

The technique of ultrasound needle guidance is essentially the same for all fetal therapy. To avoid taking the transducer out of clinical use and avoid any potential damage to the crystals, the transducer itself is not sterilized. The transducer and the cable are placed inside a gas sterilized polyethylene bag. If the bag is contaminated during the procedure, it can be replaced quickly. The face of the transducer is coated with coupling gel prior to insertion into the sterile plastic bag. Using sterilized tape, the bag is securely folded around the transducer so that a single layer of plastic (not a seam) is smoothly applied to the gel-coated face. The outside surface of the plastic bag over the transducer face is then coated with sterile coupling gel. The polyethylene is essentially transparent to ultrasound. As the transducer and cable are covered in sterile plastic, the unit can easily be used in the surgical field for transfusion. The transducer can be placed adjacent to the puncture site during needle placement. Polyethylene trash bags, masking tape, and foil packets of surgical lubricant can all be gas-sterilized together for use in these procedures.

The maternal abdomen is prepped and draped and the mother is mildly sedated but awake. A site on the maternal abdomen is then selected for the introduction of the needle. If possible, the placenta is avoided; a point overlying the anterolateral surface of the lower fetal abdomen is chosen. The fetal bladder is the most useful landmark in identifying the appropriate cephalocaudal level on the fetus for needle entry. If necessary, external version either from breech to vertex or vice versa is carried out to bring the anterolateral surface of the lower fetal abdomen into the desired position.

As the transducer is covered with sterile plastic, it can be placed directly adjacent to the point of needle insertion. By carefully keeping the needle and the transducer beam coplanar, the progress of needle insertion can be observed on the ultrasound screen. Contact with the fetal abdomen can be detected both by resistance to advancement of the needle and by indentation of the fetus on the ultrasound screen. The needle is directed toward the fetal bladder or minimally cephalad to it. This keeps the point of penetration within the lower fetal abdomen and well away from the liver and spleen, either of which may be enlarged and vulnerable to laceration. Once the fetus is contacted, the needle is advanced a few more centimeters. Penetration of the fetal abdominal wall is generally easily visualized. The exact depth of needle penetration may be difficult to ascertain by ultrasound scan. The needle occasionally appears longer on the screen than it is, because of lateral beam width artifact. It may also appear shorter if it passes out of the plane of the ultrasound beam. Ultrasonic visualization of the needle may be facilitated by sliding the stylet back and forth within the needle lumen. The hollow needle and the needle-stylet assembly have different reflective properties and the moving stylet is easily seen. This gives the operator added confidence in the displayed needle position. The presence of significant fetal ascites aids in the determination of the depth of penetration into the fetal abdomen. The needle tip can be seen clearly when surrounded by ascitic fluid (Fig. 21-1).

Once the needle appears to be in the fetal peritoneal cavity, an attempt is made to aspirate any ascites. Occasionally the needle penetrates the bladder and fetal urine is obtained. This is easily distinguished from ascites by its low protein content and lack of opalescence. If the bladder has been entered, the needle is withdrawn and reinserted slightly cephalad. If ascites is encountered, as much as possible is removed before commencing transfusion. If no fluid is obtained, additional attempts are made to confirm intraperitoneal placement of the needle. A small amount of sterile saline is injected. Intraperitoneal injection should encounter almost no resistance, and it is usually visible on the ultrasound screen because small amounts of air are displaced from the

needle. If there is no resistance to the injection, a small amount of air is injected. Bubbles are easily seen on the screen. If they rise to the highest point in the fetal abdomen and remain trapped there, intraperitoneal placement is almost assured.

Once the peritoneal placement is confirmed, infusion of blood is begun. Fetal heart rate is monitored during the infusion. Specially packed 0 Rh-negative red blood cells are used, with a hematocrit of 80 to 90%. The blood is cross-matched against that of the mother. The cells are infused through the needle in 10-ml increments by syringe. The volume of blood infused is determined according to gestational age by the formula:
volume = (gestational age in weeks − 20) × 10.
For example, at 28 weeks
 volume = (28 − 20) × 10 = 80 ml.
The entire volume is given over approximately 30 minuts. If fetal heart rate abnormalities are noted in response to the infusion, it is halted until they have resolved or the procedure is terminated.

Aggressive management of the patient with severe Rh disease has resulted in significant improvement in fetal survival rates. Some studies have reported survival rates as high as 100% for the non-hydropic fetus and 75% in the hydropic fetus. Procedure-related fetal deaths occur in less than 10% of cases. Harman et al. demonstrated that one of the most important variables in determining fetal outcome was the experience of the surgical team. Given an experienced team the severely affected fetus under 26 weeks gestational age and those with hydrops fetalis may now be suc-

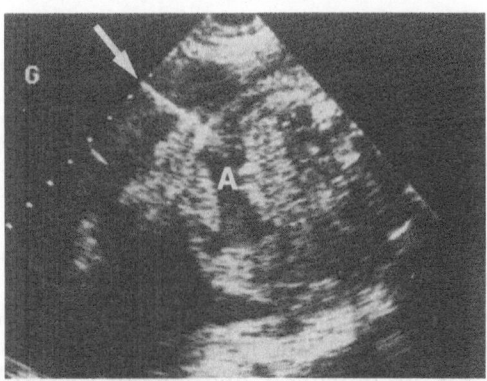

Fig. 21-1. Fetal transfusion
Real-time ultrasound exam during fetal transfusion demonstrating needle (arrow) penetrating fetal abdominal wall. Ascites (A) is present.

cessfully treated with an appropriate and aggressive approach. It is clear that referral of the severely sensitized patient to a major center with experience and skill in the treatment of this disorder is advisable.

Fetal tachycardia with congestive heart failure

Fetal cardiac arrhythmias are being detected with increasing frequency due to the proliferation of real-time equipment and the increase in electronic fetal monitoring. Persistent fetal tachycardia can result in congestive heart failure and hydrops fetalis. In any fetus with ascites, pleural effusions, or skin edema, a careful examination of the fetal heart is imperative. As fetal tachycardia may be intermittent, several measurements of the fetal heart rate and rhythm are advisable. In addition, a careful anatomic analysis of the fetal heart is necessary. While the fetus with arrhythmia usually has normal cardiac anatomy, arrhythmias have been associated with various anatomic defects including intracardiac tumors. We have detected 5 fetuses with supra-ventricular tachycardia, 4 of whom had fetal hydrops. All had evidence on fetal echocardiogram of ventricular failure. They all had a heart rate over 240 beats per minute. Four had no anatomic abnormalities. The 5th had an intracardiac tumor, presumed to be a rhabdomyoma. The mothers were digitalized and the fetal heart rate returned to normal in 3 of the cases. In the 4th, additional therapy with propranol and verapamil was tried after an initial failure of digitalis to correct the rate. When these drugs also failed, the women were treated with sustained high doses (0.75

mg/day) digoxin. After 1 week the fetal heart rate converted and the hydrops resolved. In the 5th case, with the intracardiac tumor, the rate converted with digitalis but maternal toxemia of pregnancy necessitated delivery before there was time to see resolution of the hydrops.

High dose maternal digoxin therapy has been almost universally successful in converting fetal supraventricular tachycardia. Since maternal clearance of the drug is increased in pregnancy, higher doses are needed to achieve and maintain therapeutic blood concentrations. Caution must be exercised in the treatment, however, as these doses can produce toxicity in the mother. Since free (non-protein bound) drug crosses the placenta and equilibrates with the free drug concentration in the mother, it may take several days after the mother has reached therapeutic blood concentration for the fetus to equilibrate. Intravenous administration of digoxin to mother can deliver drug to the fetus more rapidly by producing bolus effects, but again care must be taken to avoid maternal toxicity.

Urinary tract obstruction

As with many fetal anatomic problems, obstructive uropathy is a spectrum of disorders. As in children, obstruction can occur at a variety of levels in the urinary tract. Unilateral uretero-pelvic junction (UPJ) obstruction is one of the most common lesions. Depending on its severity, this lesion leads to variable degrees of unilateral hydronephrosis. If the obstruction is severe and early in gestation, it may lead to unilateral cystic renal dysplasia. Obstruction may occur less commonly at almost any other point in the urinary tract, including the uretero-vesical junction and the urethra. Ureteral obstruction is generally unilateral and, thus, at worst, results in unilateral renal damage. Rarely, it may be bilateral. Urethral obstruction, as produced by posterior urethral valves, results in bilateral dilation of the urinary tract and, if severe, may result in bilateral renal dysplasia. This can be fatal to the newborn infant. Fetal renal failure from any cause results in profound oligohydramnios and this is associated with pulmonary hypoplasia. Severely affected infants generally die of pulmonary insufficiency before their renal disease becomes a significant factor in their survival.

In cases of bilateral urinary tract obstruction due to urethral valves or bilateral UPJ obstruction, there is danger of progressive renal damage and neonatal death. Since the primary problem in these disorders is obstruction, it seems reasonable to suppose that diversion of urine around the obstruction should prevent further damage. Since urethral obstruction is more common than bilateral UPJ obstruction, the most common route of diversion should be from the bladder to the amniotic cavity. While bilateral ureterostomies have been reported in 1 case, cystostomy has been and will remain the more common and generally preferable operation. Diversion of fetal urine around urethral obstruction has been accomplished by ultrasound guided placement of shunts leading from the bladder to the amnion. In this procedure a needle is inserted into the fetal bladder and then a shunt is inserted either over or through the needle into the bladder. The shunt is then displaced from the needle as the needle is withdrawn so that the distal end is left in the amnion. This operation has been carried out in a number of centers in the U.S. and in Europe. The procedure appears to be quite feasible and many successful shunt placements have been reported. Successful outcome, as measured by neonatal survival has, however, been much less rewarding owing largely to associated pulmonary hypoplasia. Survival rates reported to date have ranged from 30% in fetuses treated at less than 20 weeks gestation, to 20% in fetuses treated after 20 weeks gestational age. The survival rate does not appear to be affected by the presence or absence of oligohydramnios or hydronephrosis.

Patient selection in obstructive uropathies is very difficult. The fetus with severe obstruction early in gestation may be doomed to cystic renal dysplasia and death. Rarely are these patients good candidates for fetal intervention. Many fetuses with moderate obstruction but who still maintain urinary output will probably survive without prenatal treatment. Between these two extremes is a group of individuals with moderate to severe obstruction and progressive renal damage who might be helped by intervention. The difficulty lies in distinguishing this group from each of the other groups. Clearly, gestational age is an important criterion. Beyond 32 weeks gestation, expectant management or premature delivery may be preferable to fetal treatment. At less mature gestational age candidates for treatment must be distinguished from the fetuses with irreversible renal damage and the fetuses with non-progressive disease. Decreasing amniotic fluid volume or oligohydramnios may distinguish the candidate from the fetus with non-progressive disease. The extent of irreversible renal damage is more difficult to detect. A two-stage experimental approach to this situation has therefore been proposed. First, fetal urinary tract catheterization with external drainage is performed in order to assess fetal urine production. While precise normal values are not known for all gestational ages, the absence of urine flow implies a poor prognosis and no further intervention may be considered. If significant urine production is present, then *in utero* urinary tract diversion should be attempted.

Clearly, treatment of fetal urinary tract obstruction by means other than premature delivery is experimental. Patient selection criteria are not well established. Risks and benefits of treatment are not fully known, and the long-term prognosis for treated and surviving infants in unknown. Because we lack precise selection criteria many treated fetuses may die of their disease and its consequences despite intervention. It is also clear, however, that some treated survivors might have survived without *in utero* treatment.

Fetal hydrocephalus

Obstructive hydrocephalus is dilatation of the cerebral ventricular system due to an increase in cerebral spinal fluid pressure and has an incidence of 0.5 to 1.8 per thousand births. The causes of congenital hydrocephalus are multiple. The majority of cases presenting at birth have no clear-cut etiology and are probably due to a combination of genetic and environmental influences (multifactoral

inheritance). A minority of cases are inherited as X linked (2%) traits or are due to autosomal recessive disorders. Fetal infection (cytomegalovirus, toxoplasmosis or rubella) may also cause hydrocephalus. The distribution of causes of hydrocephalus presenting in the fetus is unknown. With the advent of high resolution real-time ultrasound it is now possible to confidently diagnose fetal hydrocephalus prior to 18 menstrual weeks. Extreme care must be exercised, however, as a number of false positive diagnoses have been made. No therapy should be instituted without the second opinion of an experienced ultrasonographer.

Given multiple possible etiologies, it is vitally important that every effort be made to determine the specific etiology in individual cases. Appropriate investigations include careful ultrasound examination, karyotype, viral cultures and serology as well as a careful genetic family history. It is important to note, however, that many cases of congenital hydrocephalus will occur without relevant antecedent family or pregnancy history.

The cerebral spinal fluid (CSF) hypertension in congenital hydrocephalus can be caused by a number of specific mechanical problems in the brain. Complete or partial obstruction of the cerebral aquaduct is probably the most common. Communicating hydrocephalus, in which there is obstruction to absorption of cerebral spinal fluid by the subarachnoid granulations, can also cause congenital hydrocephalus. Dandy-Walker malformation and Arnold-Chiari myelodysplasia cause internal hydrocephalus due to obstruction at the level of the fourth ventricle. Fetal hydrocephalus secondary to intraventricular hemorrhage has been seen.

The prognosis for congenital hydrocephalus is quite variable and depends upon the etiology as well as the time of onset and severity of the process. The management of a pregnancy complicated by fetal hydrocephalus is made more difficult by our incomplete knowledge of its natural history. In some cases the process is relentlessly progressive throughout gestation and results in a grossly distorted fetal head and profound brain damage. In other cases hydrocephalus appears to spontaneously arrest or is only slowly progressive; occasionally it appears to resolve completely.

In recent years a number of attempts have been made to treat hydrocephalus *in utero*. The goal of these procedures has been to relieve CSF hypertension temporarily while awaiting fetal maturity in hopes of arresting progressive brain damage.

The percutaneous placement of a ventricular amniotic shunt for the relief of fetal hydrocephalus was first reported in 1982 by Clewell et al. This entailed the insertion of a needle through the maternal abdominal wall and uterus into the dilated fetal lateral ventricle under ultrasound guidance. A silastic shunt containing a one-way valve was inserted through the needle and left to extend from the ventricle to the amniotic space (Fig. 21-2). At the time of shunt placement in our 1st case, (23 weeks gestation) the LVW/HW ratio was 0.87. Following shunt placement this ratio declined to 0.50 and the cortical mantle became correspondingly thicker. After shunt placement the ventricles were asymetric with the shunt ventricle (right) consistently smaller than the left.

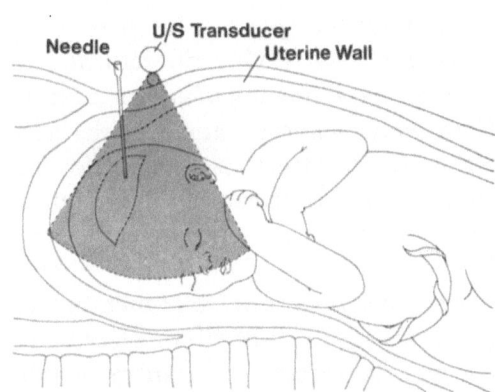

Fig. 21-2. Puncture of fetal head
Schematic diagram depicting needle placement in fetal head under ultrasound guidance. Needle should be within the plane of the ultrasound at all times.

Between 32 and 34 weeks the shunt appeared to stop functioning. The LVH/HW ratio increased and the fetal biparietal diameter was found to be 10 cm. Following delivery by Cesarean section it was found that the shunt was obstructed by an ingrowth of tissue from the ventricular end. The infant had a ventriclo-peritoneal shunt placed on the 1st day of life and was discharged at 28 days of age. This infant had a family history consistent with x-linked aquaductal stenosis. At birth he had flexion contractures of both hands and duplication of the distal phalanx of the left thumb. At 2 years, he has profound psychomotor retardation.

Four additional fetuses have been treated at the University of Colorado School of Medicine with similar procedures. None of these have had significant family histories or other apparent etiologies for the hydrocephalus. In 2 infants the shunts became displaced from the fetal head several days to weeks after placement. In both cases the ventricles rapidly enlarged again and a second shunt was placed. In these 2 infants the LVW/HW ratio returned to near normal following *in utero* shunting. Both were delivered at 32 weeks gestation by Cesarean section following spontaneous rupture of membranes. Both had ventriculo-peritoneal shunts in the neonatal period. The 3rd fetus had spontaneous rupture of membranes the day after *in utero* shunt placement. He was delivered by Cesarean section section 3 weeks later at 28 weeks gestation. He had a ventriclo-peritoneal shunt and was doing well at discharge from the nursery.

The 5th was found to have no pressure gradient from his lateral ventricle to the amnion at the time of operation. While his ventricles were clearly enlarged, the lack of significant pressure elevation suggested that drainage would be of no value and a shunt was not placed. The pregnancy was subsequently terminated. Pathologic examination of the fetal brain showed no evidence of aquaductal stenosis. Neither was there any histologic evidence of elevated intraventricular pressure. Whether this case represents an error in diagnosis or early hydrocephalus without the usual histologic evidence is unclear. In a subsequent case in which elective pregnancy termination occurred without attempted fetal treatment, definite aquaductal obstruction was found as well as vetriculomegaly, but there was no histologic evidence of elevated ventricular pressure. These cases indicate that one must interpret the histologic evidence for hydrocephalus with caution.

The postnatal development of these infants with *in utero* shunts has ranged from the severe retardation seen in the 1st case to essentially normal development in the 2nd. Given the uncertainties as to the course of fetal hydrocephalus and its multiple etiologies, such variation is not surprising. To date over 24 fetuses have been treated with similar operations in several centers. No significant maternal morbidity has been reported. There has been one procedure-related fetal death and 3 deaths due to associated congenital anomalies. While detailed developmental data are not yet available on the surviving infants, approximately 48% are considered normal neurologically, 10% mildly abnormal and 42% severely retarded, at a mean follow-up age of approximately 9 months.

The limited experience with experimental *in utero* treatment of hydrocephalus precludes any conclusions as to its efficacy at this time. It also raises several difficult questions regarding moral and ethical issues and allocation of human resources which are beyond the scope of this chapter. Several generalizations can, however, be made. Treatment with ventricular amniotic shunts can decrease ventricular size and prevent further progression of the disease. On the other hand, duration of benefit has been limited by shunt dislodgement and obstruction. Since these are both mechanical or technical problems, it seems likely that they can eventually be overcome. In cases of progressive ventriculomegaly due to increased intracranial pressure, fetal shunting can restore normal anatomy (as seen on ultrasound and computerized tomography) and result in an infant with a normal size head and normal size ventricles. Some of these infants will have completely normal neurologic and developmental exams at discharge from the nursery and at 1 year of age. Encouraging as these findings are, it is also clear that some infants will be significantly retarded even with apparently successful treatment. The developmental prognosis probably depends as much on the etiology of the hydrocephalus as it does on the severity or stage of pregnancy when the shunting is performed.

In selecting patients for fetal shunt placement several criteria must be met. First, other serious anomalies should be excluded as thoroughly as possible. This requires a very meticulous ultrasound examination, utilizing the most sophisticated equipment and interpretation. Amniocentesis for karotype and alpha fetal protein should be performed. Second, the procedure should probably not be performed beyond 32 weeks gestation since neonatal survival is quite good in modern nurseries at that stage. Third, consultation involving obstetricians, ultrasonologists, pediatricians, geneticists, pediatric neurosurgeons and psycho-social services must be obtained. Lastly and most importantly, the family must be completely informed of the risks, benefits and experimental nature of the procedure before being asked to consent to it.

Fetal shunting for hydrocephalus is and will remain for some time to come an experimental procedure. Criteria for patient selection, operative technique, and equipment continue to evolve. It will be several years before meaningful follow-up data are available to assess the benefit of this procedure. Extraordinary caution in patient selection and follow-up is mandatory during this experimental period.

References

Harrison M R, Golbus M S, Filly R A. Management of the fetus with a correctable congenital defect. *JAMA* 1981; 246: 774.

Liley A W. Intrauterine transfusion of foetus in haemolytic disease. *Br Med J* 1963; 2: 1107.

Acker D, Frigoletto F E, Birnholz J C, et al. Ultrasound-facilitated intrauterine transfusions. *Am J Obstet Gynecol* 1980; 138: 1200.

Berkowitz R I, Hobbins J C. Intrauterine transfusion utilizing ultrasound. *Obstet Gynecol* 1981; 57: 33.

Clewell W H, Dunne M G, Johnson M L, et al. Fetal transfusion with realtime ultrasound guidance. *Obstet Gynecol* 1981; 57: 516.

Larkin R M, Knochel J Q, Lee T G. Intrauterine transfusions: new techniques and results. *Clin Obstet Gynecol* 1982; 25: 303.

Harman C R, Manning F A, Bowman J R, et al. Severe RH disease-poor outcome is not inevitable. *Am J Obstet Gynecol* 1983; 145: 823.

Harrigan J T, Kangos J J, Sikka A, et al. Successful treatment of fetal congestive heart failure secondary to tachycardia. *N Engl J Med* 1981; 304: 1527.

Wiggins J W, Clewell W, Johnson M L, et al. Successful diagnosis and therapy of arrhythmias, congestive heart failure in the fetus. Submitted for Publication.

Harrison M R, Filly R A, Parer J T, et al. Management of the fetus with a urinary tract malfunction. *JAMA* 1981; 246: 635.

Harrison M R, Golbus M S, Filly R A, et al. Fetal surgery for congenital hydronephrosis. *N Engl J Med* 1982; 306: 591.

Golbus M S, Harrison M R, Filly R A, et al. In utero treatment of urinary tract obstruction. *Am J Obstet Gynecol* 1982; 142: 383.

Blane C E, Koff S A, Bowerman R A, et al. Nonobstructive fetal hydronephrosis: sonographic recognition and theraputic implication. *RAD* 1983; 147: 95.

Manning F A, Harman C R, Lange I R, et al. Antepartum chronic fetal vesicoamniotic shunts for obstructive uropathy: a report of two cases. *Am J Obstet Gynecol* 1983; 145: 819.

Kramer S A. Current status of fetal intervention for congenital hydronephrosis. *Urol* 1983; 130: 641.

Robertson R D, Sarti D A, Brown W H, et al. Congenital hydrocephalus in two pregnancies following the birth of a child with neural tube defect: etiology and management. *J Med Genet* 1981; 18: 105.

Bay C, Kerzin L, Hall B. Recurrence risk in hydrocephalus. Birth Defects: Original Article Series 1979; 15(5C): 95.

Habib Z. Genetics and genetic counselling in neonatal hydrocephalus. *Obstetrical and Gynecological Survey* 1981; 36: 529.

Johnson M L, Dunne M G, Mack L A, et al. Evaluation of fetal intracranial anatomy by static and real-time ultrasound. *J Clin Ultrasound* 1980; 8: 311.

Hadlock F P, Deter R L, Park S K. Real-time sonography: ventricular and vascular anatomy of the fetal brain in utero. *AJR* 1981; 136: 133.

Denkhaus H, Winsberg F. Ultrasonic measurement of the fetal ventricular system. *Radiology* 1979; 131: 781.

Kim M S, Elyaderani M K. Sonographic diagnosis of cerebroventricular hemorrhage in utero. *Radiology* 1982; 142: 479.

McGahan J P, Haessloin H C, Meyer M, et al. Sonographic recognition of in utero interventricular hemorrhage. *AJR* 1984; 142: 171.

Machejda M, Hodgen J D. In utero diagnosis and treatment of non-human primate fetal skeletal anomalies. I. Hydrocephalus. *JAMA* 1981; 246: 1093.

Birnholz J C, Frigoletto F D. Antenatal treatment of hydrocephalus. *N Engl J Med* 1981; 304: 1021.

Clewell W H, Johnson M L, Meier P R, et al. A surgical approach to the treatment of fetal hydrocephalus. *N Engl J Med* 1982; 306: 1320.

Elias S, Annas G. Perspectives on fetal surgery. *Am J Obstet Gynecol* 1983; 145: 807.

By kind permission this chapter is taken from: The Principles and Practice of Ultrasonography in Obstetrics and Gynecology, 3rd edition, edited by Roger Sanders, M D and A Everette James, Jr., M D. Copyright 1985, Appleton – Century – Grofts, East Norwalk, C T.

Puncture of retroperitoneal mass lesions

Niels Juul, Søren Torp-Pedersen & Bo Hainau

Retroperitoneal masses are either primary or secondary. Primary lesions are sarcomas, carcinomas or benign neoplasms or lesions due to diseases of the lymphoreticular system involving the retroperitoneal nodes. Secondary tumors are metastases to the lymph nodes from a wide variety of primary sites.

When a retroperitoneal mass is suspected, several diagnostic modalities may be applied. CT and NMR provide excellent images of the retroperitoneum, providing sections which are not impeded by air or fat. Lymphography gives information about the contrast uptake of the iliac and aortic lymph nodes and tumors in the lymph nodes are visualized either directly by a pathological contrast medium uptake or indirectly by the absense of uptake or displacement of normal lymph nodes. This modality is generally most reliable in the lower retroperitoneal space whereas visualization of the upper retroperitoneal nodes may be difficult.

With ultrasound, overlying intestinal gas may spoil the image. However, by pressing the transducer firmly against the abdomen the problem is usually solved. In that respect a dynamic sector scanner (mechanical, phased array or curved array) is preferable. Further-

more, the left liver lobe is an excellent acoustic window for the superior part of the retroperitoneal space and the urinary bladder for the inferior part.

The great variety in retroperitoneal lesions is reflected in the ultrasound images. Carcinomatous metastases are often irregular with bleeding and necrosis and most frequently appear as irregular relatively echo-rich masses (Fig. 22–1). Lymphomas, being very homogeneous tumors, appear echo-poor, often almost echo-free, on the ultrasound image (Fig. 22-2). However, it is not possible by ultrasound or by any other diagnostic modality with certainty to reveal the true nature of the mass lesion. For this evaluation a biopsy is necessary.

Fluoroscopic needle guidance requires radiopacity of the target or of the surroundings; CT guidance does not require radiopacity, but apart from slight angulations, it is limited to the transverse plane. Needle guidance by either modality is time-consuming and cumbersome.

With dynamic ultrasound, any structure visualized can easily and quickly be reached by a fine needle in any desired plane with constant needle tip visualization during its insertion.

No preparation of the patient is

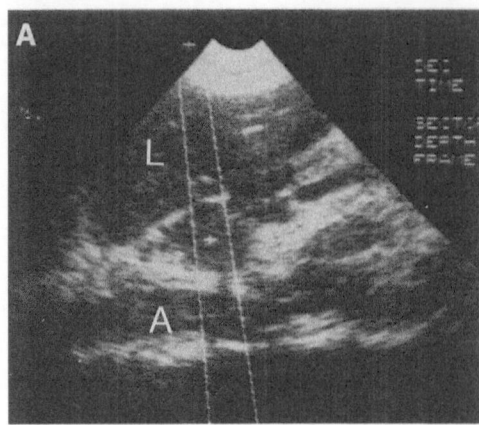

Fig. 22-1. Metastases from testicular seminoma

A: Longitudinal section of upper retroperitoneum showing an echo-rich mass. Needle path indicated. Liver: L, Aorta: A.

B: Corresponding fine needle aspirate showing malignant cells consistent with seminoma of the testis.

C: Operative specimen showing an irregular tumor with necrosis and hemorrhage (asterisk).

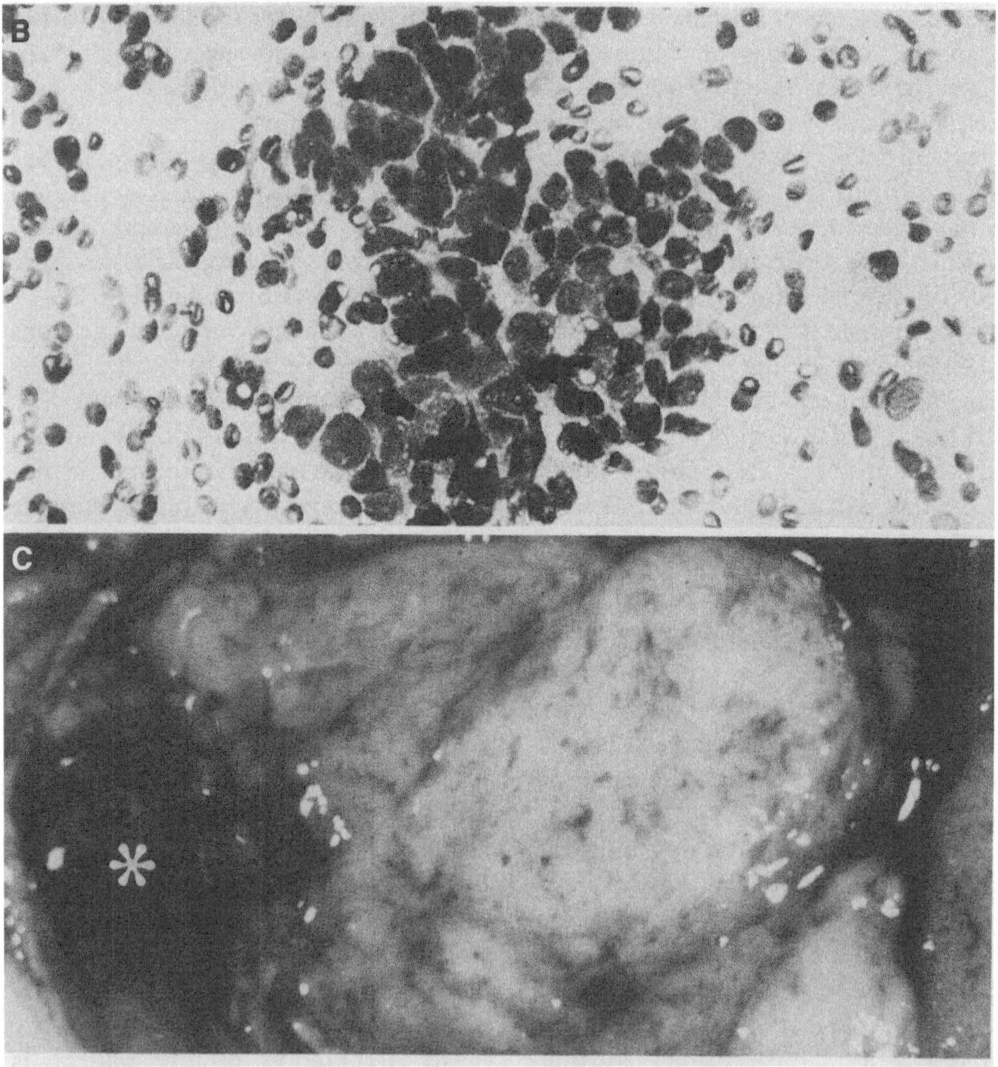

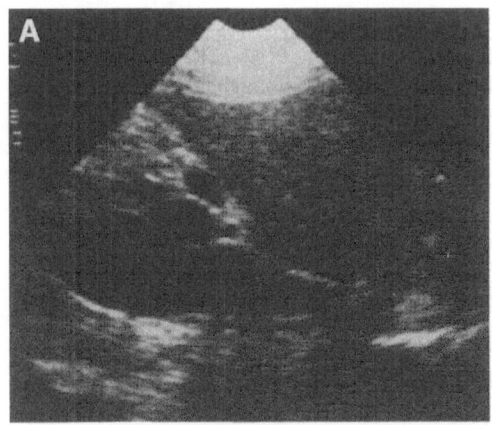

Fig. 22-2. Hodgkin's disease
A: Longitudinal section of upper retroperitoneum showing an echo-poor mass. Fine needle biopsy revealed a solid mass.
B: Corresponding fine needle aspirate showing lymphocytes and a Reed-Sternberg cell (arrow). Cytologic diagnosis: Hodgkin's disease.
C: Hodgkin's disease (operation specimen). Note the homogenous rubbery cut surface.

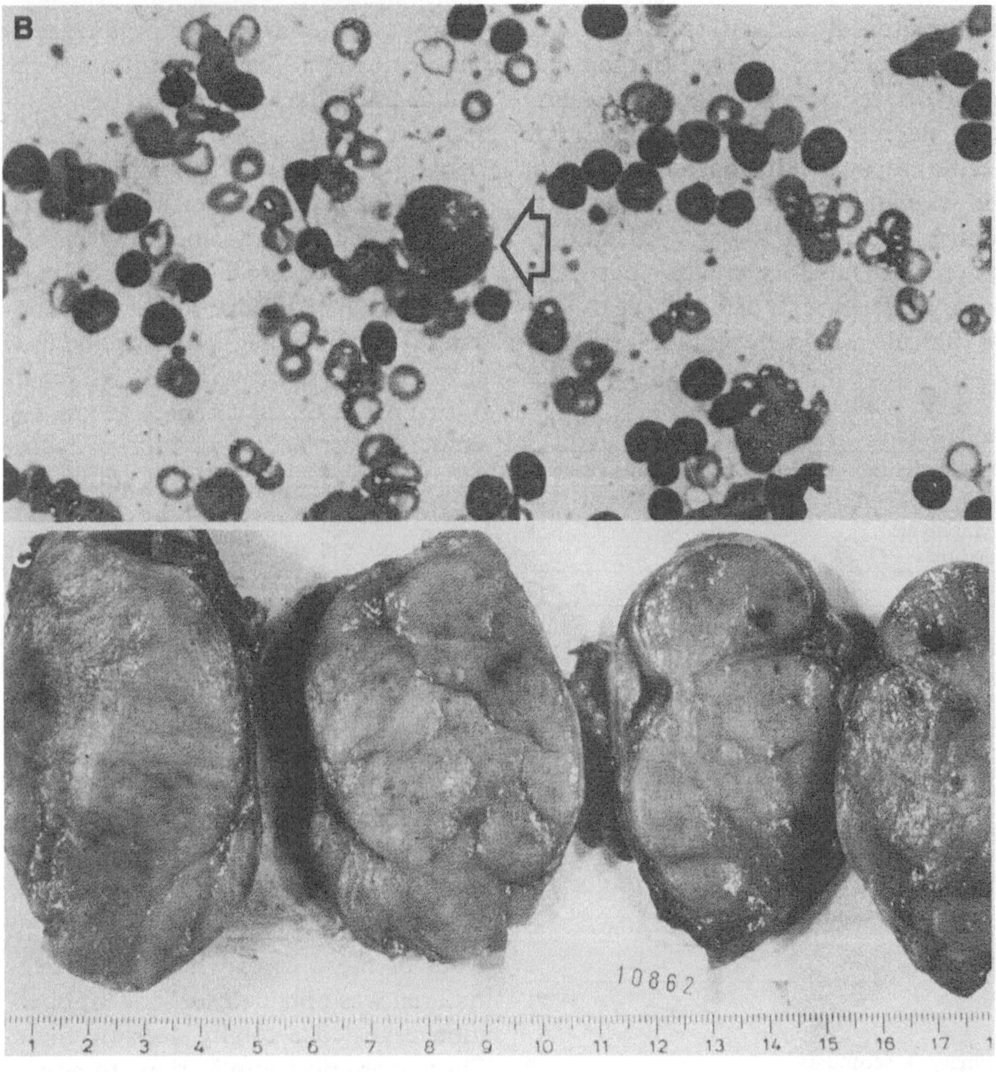

necessary and the puncture can be performed on an out-patient basis. The most convenient puncture route is chosen traversing the intestine, the urinary bladder or, if necessary, even the liver, but always avoiding the pleura and the great vessels. The puncture technique and handling of the aspirated material is described in chapters 2, 3 and 4.

Material and results

Experience with ultrasonically guided retroperitoneal biopsy has been described by several authors. In their materials, as well as in the materials dealing with fluoroscopic or CT guided fine needle biopsy the diagnostic accuracy of the imaging modalities is increased by adding the needle biopsy. All authors experience false negatives, but false positives are rare. During a 5-year period 100 ultrasonically guided retroper-

Table 22-1.

Distribution of malignant retroperitoneal lesions:	
Primary:	
Lymphomas	9
Sarcomas	8
Total	17
Secondary to:	
Liver	2
Bile radicles	2
Pancreas	7
Gastrointestinal tract	8
Lung	3
Breast	1
Kidney	11
Bladder & Prostate	7
Testes	10
Uterus and Ovaries	7
Melanoma	1
Unknown	2
Total	61
Primary + secondary	78

Table 22-2. Cytological diagnosis

Final diagnosis	Malignancy	Non-malignancy	Insufficient material
malignant 78	67	6	5
benign 22	0	21	1
total 100	67	27	6

predictive value of malignant aspirate: = 100% (94–100).

predictive value of non-malignant aspirate: = 78% (58–91).

itoneal biopsies have been performed at our institution. The distribution of the malignant lesions are listed in Table 22-1 and the results shown in Table 22-2. Of the 100 punctures, material sufficient for cytological evaluation was obtained in 94. A correct cytological diagnosis was established in 88 cases. In 78 patients with a malignant lesion, a malignant cytological diagnosis was obtained in 67 giving a sensitivity of 85%. Twenty-two patients had a benign lesion (i.e. simple hyperplasia of a lymph node, inflammatory changes, fibrosis and cysts). There were no false positives.

The means of verification were: for the malignant lesions in 48 cases autopsy or operation and in 30 cases compelling clinical evidence (i.e. similar findings on CT, angiography, lymphography etc., or malignant course of the disease); for the benign lesions in 13 cases operation and in 9 cases compelling clinical evidence (other diagnostic modalities in agreement with the diagnosis and benign course of the disease, the observation time being at least 1 year). In 4 out of the 6 cases where insufficient material was obtained subsequent operation revealed a totally necrotic mass. There were no complications to any of the biopsies in our material.

In spite of several investigations stating that ultrasound is as reliable a modality as CT and probably superior to lymphography, there is still a strong tendency to regard CT as the best of the three. However, the value of ultrasound is probably underestimated because too little credit is given to its easy combination with accurate fine needle aspiration biopsy, enabling direct needle monitoring.

When evaluating the usefulness of ultrasound in retroperitoneal diagnosis the guided fine needle aspiration biopsy should be regarded as an integral part of the examination, as it requires no preparation of the patient and is a rapid procedure with a high diagnostic sensitivity and specificity.

References

Göthlin J H. Post-lymphographic percutaneous fine needle biopsy of lymph nodes guided by fluoroscopy. *Radiology* 1976; 120: 205.

Pereiras R V, Meiers W, Kunhardt B et al. Fluoroscopically guided thin needle aspiration biopsy of the abdomen and retroperitoneum. *AJR* 1978; 131: 197.

Macintosh P K, Thomson K R, Barbaric Z L. Percutaneous transperitoneal lymph node biopsy as a means of improving lymphographic diagnosis. *Radiology* 1979; 131: 647.

Zornoza J, Wallace S, Goldstein H M, Lukenman J M, Jung B. Transperitoneal percutaneous retroperitoneal lymph node aspiration biopsy. *Radiology* 1977; 122: 111.

Haage J R. New techniques for CT-guided biopsies. *AJR* 1979; 133: 633.

Sundaram M, Wolverson M K, Heiberg E, Pilla T, Vas W G, Shields J B. Utility of CT-guided abdominal aspiration procedures. *AJR* 1982; 139: 1111.

Zornoza J, Cabanillas F F, Altoff T M, Ordonez N, Cohen M A. Percutaneous needle biopsy in abdominal lymphoma. *AJR* 1981; 136: 97.

Pedersen J F. Percutaneous puncture guided by ultrasonic multitransducer scanning. *J Clin Ultrasound* 1977; 5: 175.

Holm H H, Hancke S, Grønvall S, Jacobsen G K. Interventional ultrasound. In: Proceedings from the Third Meeting of the World Federation for ultrasound in Medicine and Biology. Oxford, New York: Pergamon, 1984.

Lindgren P G. Ultrasonically guided punctures. A modified technique. *Radiology* 1980; 137: 235.

Porter B, Karp W, Forsberg L. Percutaneous cytodiagnosis of abdominal masses by ultrasound guided fine needle aspiration biopsy. *Acta Radiol* (Diagnosis) 1981; 22: 663.

Juul N, Torp-Pedersen S, Holm H H. Ultrasonically guided fine needle aspiration biopsy of retroperitoneal mass lesions. *Br J Radiol* 1984; 45: 43.

Rochester D, Bowie J D, Kunzmann A, Lester E. Ultrasound in staging of lymphomas. *Radiology* 1977; 124: 483.

Burney B T, Klatte E C. Ultrasound and computed tomography of the abdomen in the staging and management of testicular carcinoma. *Radiology* 1979; 132: 415.

Brascho D J, Durant J R, Green L E. The accuracy of retroperitoneal ultrasonography in Hodgkin's disease and non-Hodgkin's lymphoma. *Radiology* 1977; 125: 485.

Hutschenreiter G, Alken P, Schneider H M. The value of sonography and lymphography in the detection of retroperitoneal metastases in testicular tumors. *J Urol* 1979; 122: 766.

Puncture of gastrointestinal mass lesions

Søren Torp Pedersen, Niels Juul, Torben Larsen, Hans Henrik Holm & Maxwell Sehested

Lesions of the gastrointestinal tract are usually diagnosed by means of endoscopy and barium studies. Ultrasound is generally not considered useful in gastrointestinal diagnostics because of its inability to penetrate gas. However, as already described in 1976, by pressing the transducer firmly against the abdomen intestinal gas will be displaced and tumors may be disclosed.

A tumor of the gastrointestinal tract presents itself on the ultrasound image as the so-called "bull's-eye" or "target lesion". It consists typically of a rounded, echo-poor, solid structure with a central echo-rich core when sectioned transversely. The echo-rich core is the lumen and the surrounding echo-poor region the thickened wall. When sectioned longitudinally the echo-rich core is seen as an irregular white line passing through an oblong tumor (Figs. 23-1, 23-2, & 23-3).

It is important that the lesion is observed for some time using dynamic scanning. Localized contraction of a bowel loop will have the above mentioned characteristics and only the eventual peristaltic movements will distinguish it from a tumor. Furthermore, it should be kept in mind that the normal pylorus of the stomach may present

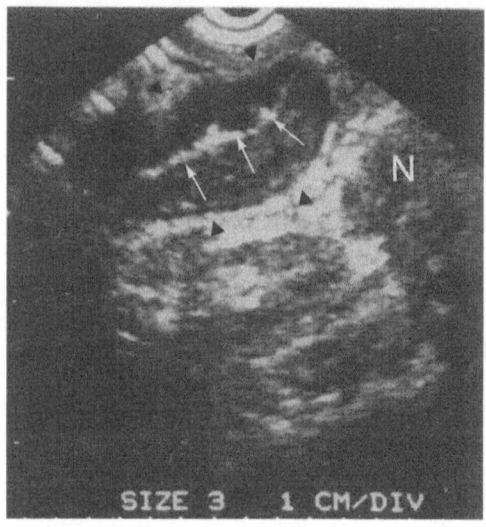

Fig. 23-1. Tumor of the colon
The tumor which is sectioned longitudinally is outlined by arrow heads and the lumen is indicated by arrows. Normal bowel loops are seen behind tumor (N).

itself as a "bull's-eye" lesion.

It is only occasionally possible from the ultrasound image to tell whether the lesion is located in the stomach, small intestine, or colon. In case of a tumor in the upper left part of the abdomen administration of water perorally will tell whether or not the lesion is located in the stomach. Lesions found along the path of the colon are, of course, suspected to be colonic tumors, but lesions

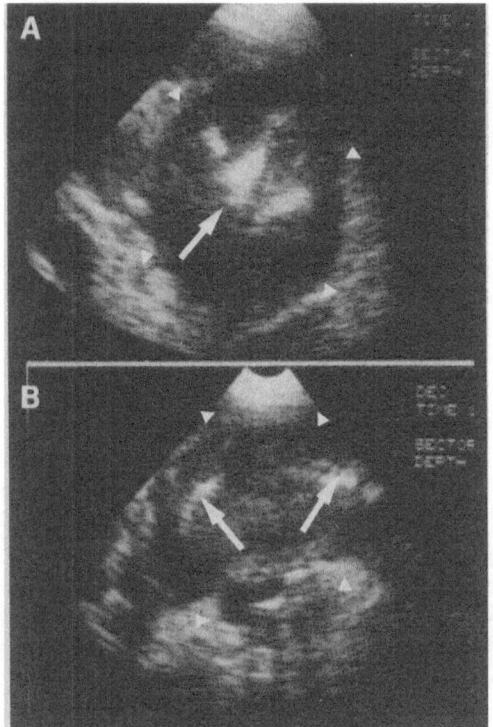

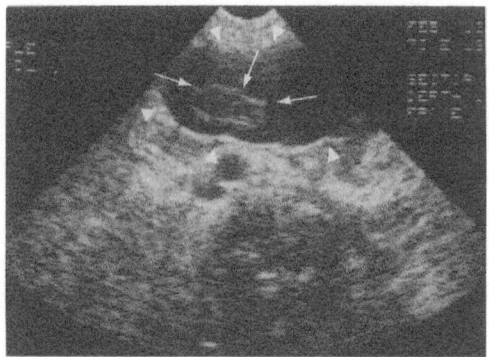

Fig. 23-3. Lymphoma of the small intestine
An echo-poor tumor (arrow heads) circumscribes the lumen (arrows) which is unaffected. Biopsy revealed non-Hodgkin lymphoma.

Fig. 23-2. Tumor of the stomach
A: Tumor sectioned transversely and B: longitudinally. The tumor is outlined by arrow heads and the luminal echoes are indicated by arrows.

The technique of fine needle aspiration biopsy of the gastrointestinal tract does not differ from aspiration biopsies from other abdominal organs. The target is often superficially located and easy to reach. The lumen of the lesion should be avoided – not for fear of compli-

of the small bowel can easily cause confusion.

The "bull's-eye" lesion merely indicates that a part of the gastrointestinal tract has a thickened wall – it does not indicate whether the lesion is malignant or not. An inflammatory condition, as for example Crohn's disease, is indistinguishable from a malignant condition (Fig. 23-4).

As in other cases of suspected abdominal malignancy the fine needle aspiration biopsy should be regarded an integral part of the examination since it does not prolong it by more than 10–15 minutes and the correct diagnosis will often be reached in one setting.

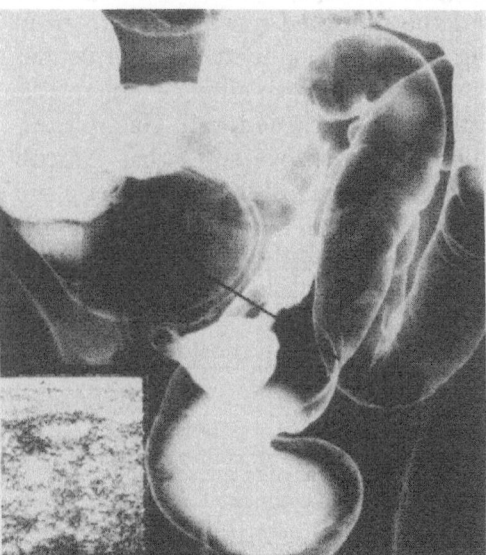

Fig. 23-4. Crohn's disease
Barium enema of the colon showing strictures of the terminal ileum. The black line indicates the scanning plane and the inset shows the ultrasonic image of the affected bowel loop.

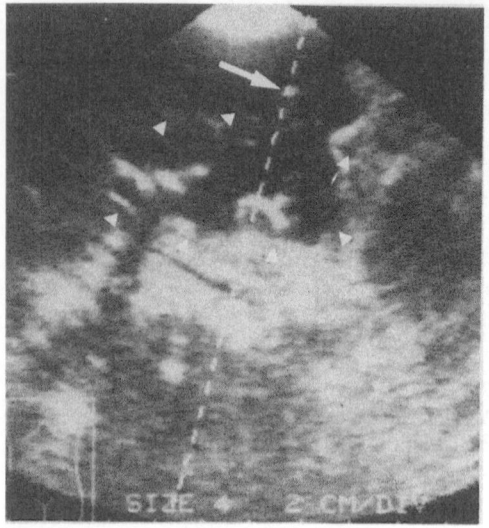

Fig. 23-5. Fine needle biopsy of a colonic tumor
The tumor is indicated by arrow heads. The tip of the fine needle (arrow) is seen as it moves along the puncture line.

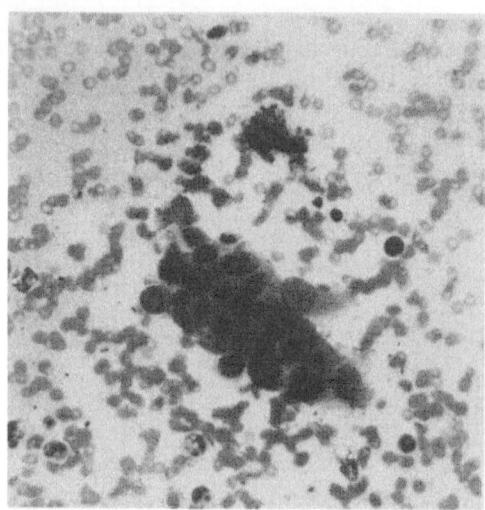

Fig. 23-6. Colonic tumor cells
Material from fine needle aspiration shown in Fig. 23-5. A cluster of tumor cells is seen on a background of erythrocytes.

cations, but because bowel juice will be aspirated instead of cells. An ultrasound image of a biopsy is shown in Fig. 23-5 and the aspirated material is shown in Fig. 23-6. The puncture route is tangential to the lumen giving a long needle path in the thickened wall. Gastrointestinal biopsies prove somewhat difficult because the tumor tends to move up and down with the needle, wherefore the movements of the needle must be exaggerated to ensure adequate sampling.

The gross appearance of the cytological material must be carefully viewed in order to ensure that cells have been aspirated and not just bowel juice.

Results

During a 6 year period we have performed 78 biopsies of ultrasonically visualized gastrointestinal tumors. The results are shown in Tables 23-1 and 23-2. Nineteen were located in the stomach,

Table 23-1. Gastrointestinal lesions

| | | Cytological diagnosis | |
	Malignant aspirate	Non-malignant aspirate	Insufficient material
Malignant solid mass 61	50	8	3
Benign condition 17	0	16	1
	50	24	4

Predictive value of malignancy $\dfrac{50 \times 100}{50}\% = 100\%$ (93–100).

Predictive value of non-malignancy $\dfrac{16 \times 100}{24}\% = 67\%$ (45–84).

95% security limits in brackets.

Table 23-2. Cytology from 78 gastrointestinal lesions

		Final diagnosis		Cytology		
Stomach	19	malignant	18	14	2	2
		benign	1		1	
Small intestine	3	malignant	3	2		1
		benign	0			
Colon	56	malignant	40	34	6	
		benign	16		15	1

Table 23-3. Benign lesions diagnosed by fine needle aspiration biopsy in 78 cases

	Number
Diverticulitis	6
Periappendicular abscess	5
Uncertain*	6
	17

*Symptoms and lesions disappeared without surgical intervention.

3 in the small intestine and 56 in the colon. In 50 of the 61 cases of malignancy the cytologic diagnosis was positive for cancer, giving a sensitivity as concerns malignancy of 82%. There were no false positives. The overall diagnostic accuracy was 85%.

Our material shows that the combination of ultrasound scanning and fine needle aspiration biopsy gives a high degree of accuracy in the diagnosis of gastrointestinal tumors. That ultrasound definitely has a role to play in gastrointestinal diagnostics is indicated by the fact that approximately half of the tumors were found primarily on the ultrasound scan.

Both endoscopy and barium studies can detect tumors only if they affect the lumen. In cases of wall infiltrating cancers the lumen may be unaffected and the tumor may therefore go undetected in some cases with both diagnostic procedures. In fact, 7 of the 40 malignant tumors of the colon in our series were not disclosed by barium enema studies (Fig. 23-7). It should be kept in mind, however, that this is a selected material and it does not show how many tumors ultrasound has overlooked in the same period.

The ultrasonically guided fine needle aspiration biopsy can also be regarded as an alternative to repeated endoscopically guided biopsies. It is well-known that endoscopically guided biopsies also yield false-negative results, especially in cases of cancers infiltrating the submucous layers. If the biopsy and therefore also the endoscopy are to be repeated, an ultrasound study should be considered as an alternative.

A wall infiltrating cancer is very likely to be disclosed at sonography and the biopsy procedure requires no preparation of the patient, as does endoscopy. In 5 of the cases of gastric malignancy in our series endoscopically guided biopsy did not yield sufficient material – in 2 cases despite repeated attempts.

In 17 cases there were benign changes and the cytology was correctly negative (Table 23-3). However, in 6 of these 17 cases, the tumor could not be found on repeated scans (Fig. 23-8). Either some normal configuration had been mistaken for a tumor or an inflammatory process had disappeared. Obvious target lesions should be punctured right away while doubtful lesions should be rescanned after a couple of days and puncture performed only when the lesion is reproducible.

That gastrointestinal fine needle puncture is safe is indicated by the complete absence of complications in this material as well as in materials describing punctures of retroperitoneally lo-

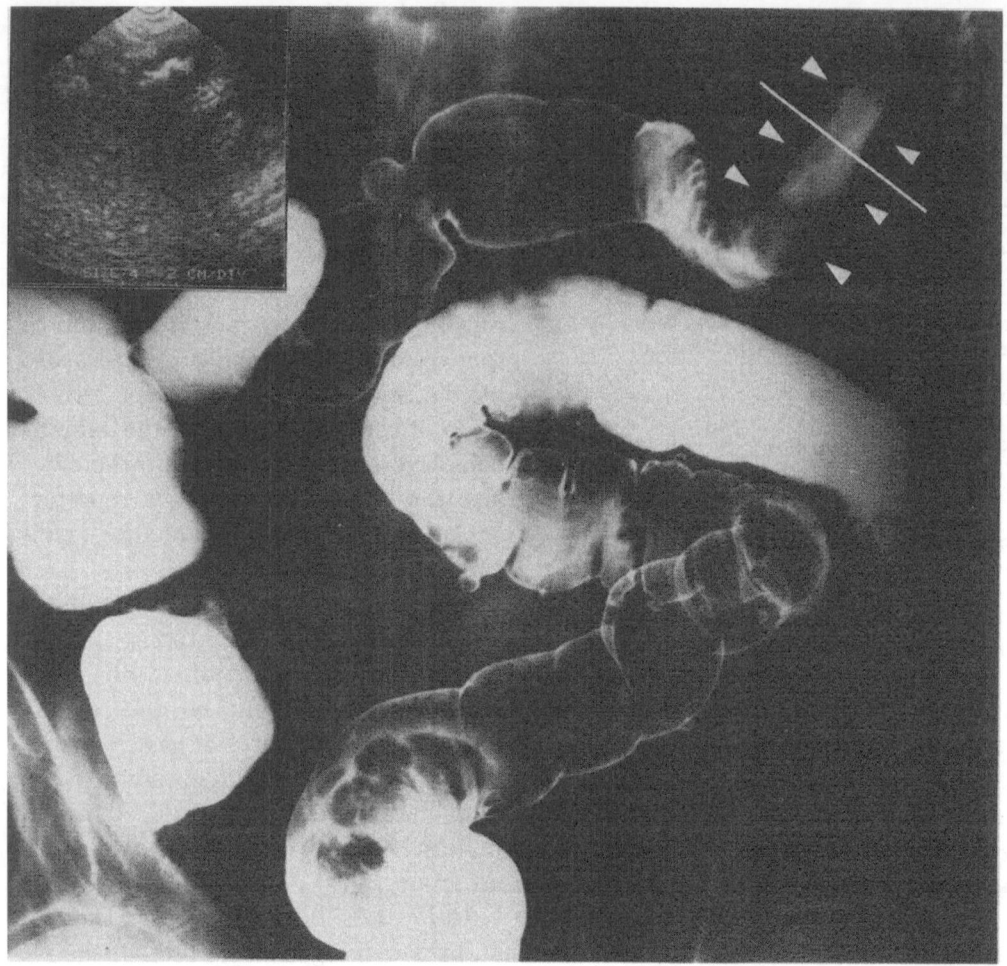

Fig. 23-7. Barium enema with tumor of the transverse colon
The study was misinterpretated as a spastic left flexure. Inset shows typical gastrointestinal tumor and the scanning plane is indicated by white line.

cated structures, where the needle traverses the gastrointestinal tract.

We feel therefore that the indications for an ultrasound study including biopsy are equivocal barium studies, unsuspected negative endoscopic biopsy, colonic structure of undetermined nature and finally a biopsy should be carried out on incidentally disclosed target lesions.

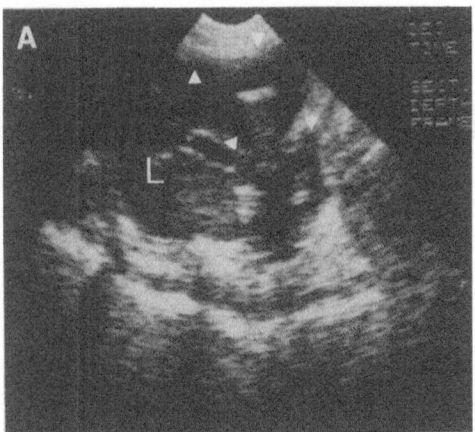

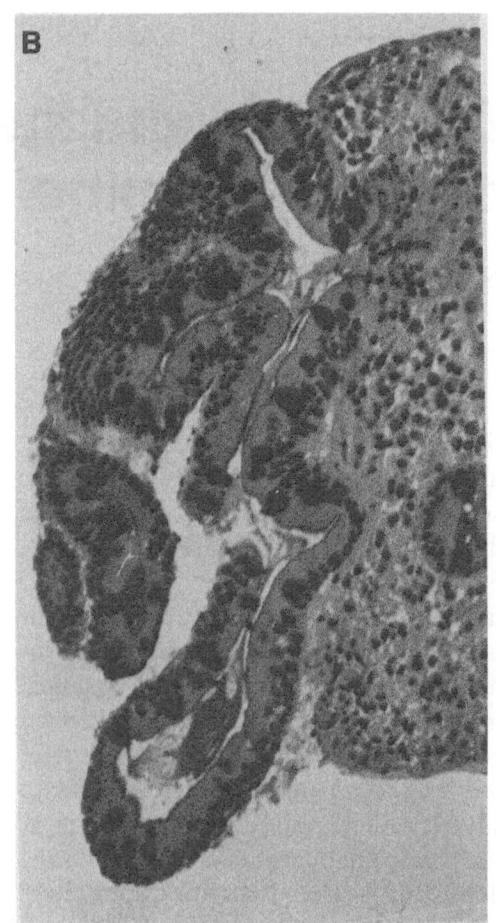

Fig. 23-8. Misinterpreted bulls-eye lesion
A: "Bull's-eye" lesion (arrow heads) below the liver (*L*) in a patient with pain below the right costal margin. Biopsy revealed no malignancy and the lesion could not be found on repeated scans. B: Material from fine needle histological biopsy showing normal villus from the small intestine.

References

Lutz H H, Petzholdt R. Ultrasonic patterns of space occupying lesions of the stomach and the intestine. *Ultrasound Med Biol* 1976; 2: 129.

Walls W J. The evaluation of malignant gastric neoplasms by ultrasonic B-scanning. *Radiology* 1976; 118: 159.

Gammelgaard J, Holm H H, Henriksen O B. Ultrasonically guided fine needle aspiration biopsy of gastrointestinal mass lesions. Abstract: 2nd Meeting of The World Federation for Ultrasound in Medicine and Biology, Miyazaki, Japan, 1979, p. 281.

Torp-Pedersen S, Grønvall S, Holm H H. Ultrasonically guided fine-needle aspiration biopsy of gastrointestinal mass lesions. *J Ultrasound Med* 1984; 3: 65.

Diagnostic and therapeutic puncture of intraabdominal fluid collections

Sven Grønvall

Sonography has become one of the most important methods in the localization and identification of intraabdominal fluid collections and abscesses. A combination of sonography and percutaneous diagnostic puncture in particular serves as an excellent sampling system providing important clinical information and material for further evaluation (see Chapter 4). Moreover, the diagnostic puncture often leads to an immediate percutaneous drainage, which especially in the treatment of abdominal abscesses is an important alternative to conventional surgery.

Dynamic scanning in particular is well suited for screening for abdominal fluid collections, the distribution of which usually follows the interconnected compartments in the peritoneal cavity (Fig. 24-1). This means that the entire abdominal cavity should be examined when a fluid collection is suspected. The shift of free fluid to a dependent localization can be demonstrated, and fluid filled intestinal loops and the stomach can be distinguished from extraluminal collections of fluid by the detection of peristalsis.

An abdominal fluid collection appears sonographically sometimes as a rounded and sometimes as a more irregularly

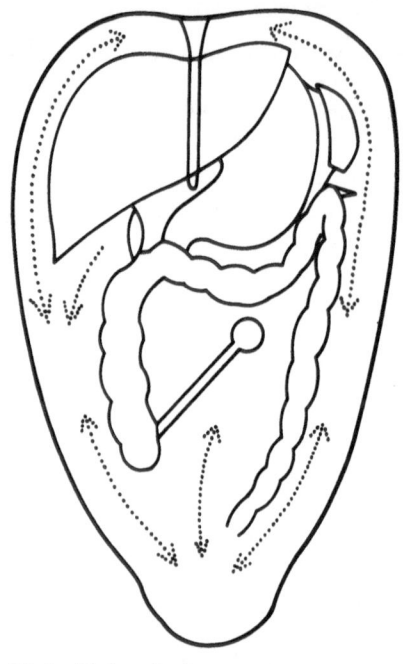

Fig. 24-1. Abdominal compartments
The different interconnected abdominal compartments. Arrows indicate some of the main routes of fluid movements.

outlined echo-free or echo-poor structure which may distort and displace adjacent organs. The internal echoes in abscesses and hematomas are caused by fibrin strands, detritus or blood clots. Frequently an abscess appears rather echogenic, suggesting a solid tumor. Often, however, the contents will move slowly following a change in the pati-

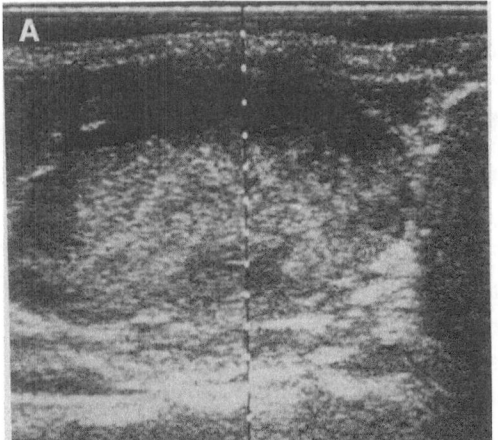

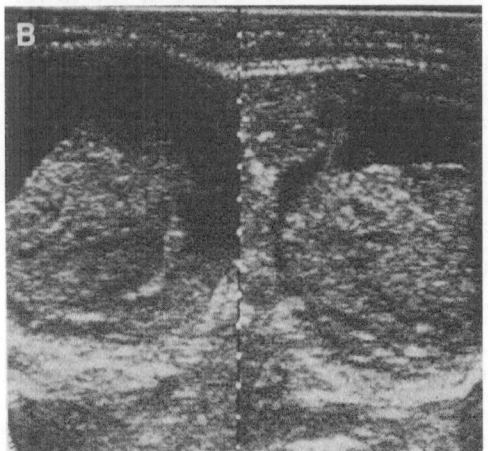

Fig. 24-2. Gall bladder empyema
A: Longitudinal scan. B: Transverse scans. A large gall bladder empyema showing a fluid-pus level that changes according to the patient's position. The empyema was catheter-drained temporarily guided by sonography before elective cholecystectomy was performed.

ent's position (Fig. 24-2.). Very strong echoes with shadowing are seen in gas-containing abscesses.

Diagnostic puncture

Precise needle placement in a fluid collection or abscess is achieved according to the principles described in Chapter 2 using a puncture transducer with a

Table 24-1.

Subphrenic	44
Intrahepatic	25
Subhepatic	52
Lower quadrants	87
Adjacent renal graft	42
True pelvis	30
Interintestinal	21
Retroperitoneal	42
Renal	26
Abdominal wall	28
Pleural cavity	7
Total	**404**

Localization of intraabdominal fluid collections punctured for diagnostic purposes.

steering device. Percutaneous puncture is an easy and almost atraumatic procedure that can be performed virtually without risk even in the critically ill patient.

In a series of 404 patients with suspected intraabdominal fluid collections, sonographically guided aspiration was performed for diagnostic purposes. Material was obtained from different abdominal compartments, the pleural cavity and various parenchymal organs (Table 24-1). The size of the fluid collections varied from a few cc's to more than 2000 cc's. The character of the aspirated fluid (Table 24-2) suggested different kinds of collections (abscess, hematoma, ascites etc.). It appears that approximately 50% were abscesses.

Table 24-2.

Pus	187
Blood	96
Ascitic fluid	60
Lymph	23
Bile	6
Urine	6
Pleural exudate	7
Undefined	11
No fluid	8
Total	**404**

Character of aspirated fluid.

155

Therapeutic puncture

As mentioned, the ultrasonically guided percutaneous puncture may serve not only a diagnostic but a therapeutic purpose. An old proverb says: "Ubi pus, ibi evacuat*". This task has always been met by conventional surgical treatment, which has few, but important drawbacks: It is distressing to the patient, time-consuming and it inevitably contaminates the hospital environment. To reduce these drawbacks the following procedure has been used for the past 6 years at The Ultrasonic Department, Herlev Hospital, Copenhagen, when a diagnostic puncture has revealed pus from a fluid collection: The pus is aspirated with an 18 gauge (1.2 mm) needle and sent for aerobic and anaerobic cultivations. The abscess is emptied completely if possible. This is often facilitated by repeated irrigations with saline. Ampicillin 1 g and cephaloridin 1 g in 10 cc sterile water is injected. Control scanning is performed after 3 days and the procedure repeated if necessary. Depending on the size and location of the abscess, a polyethylene catheter (gauge 12–14) is introduced, guided by sonography using the Seldinger technique. The catheter is left in place for drainage, subsequent irrigation and daily instillations of antibiotics.

When the scan shows disappearance of the abscess and the discharge from the cavity has ceased, the catheter is removed. To show the position of the catheter and to visualize the morphology of the cavity, X-rays can be obtained after injection of a contrast medium. Needle puncture of abscesses

*"Where there is pus, let it out."

through the gastrointestinal tract and the urinary bladder is performed without hesitation, but catheters are not inserted in these situations.

Results

A series of 108 visceral, intraperitoneal or retroperitoneal abscesses has been treated according to the principles mentioned. The location of the abscesses is shown in Table 24-3. The perihepatic and retroperitoneal groups include subphrenic, subhepatic, pancreatic and renal abscesses. Abscesses in the lower quadrants are mainly of the periappendicular type and true pelvis include Douglas and gynecologic abscesses. Forty-seven abscesses were treated with one or more punctures and 61 drained by means of a catheter (Table 24-4). Forty-two and 50, respectively, were succesfully treated giving a total curability of 85% (Fig. 24-3 and 24-4).

In 15% healing of the abscess could not be accomplished by drainage alone (Table 24-5). In general most of these patients had multi-compartment abscesses or abscesses originating from a

Table 24-3.

Perihepatic	31
Intrahepatic	18
Retroperitoneal	20
Lower quadrants	25
True pelvis	10
Abdominal wall	4
Total	108

Localization of abscesses treated by puncture and antibiotics locally.

Table 24-4.

		Cured	Non-cured
Puncture	47	42	5
Catheter	61	50	11
Total	108	92(85%)	16(15%)

Therapeutic results.

Table 24-5.

Puncture (5):	Adjacent renal graft
	Subhepatic
	Tubo-ovarian
	Adjacent necrotic intestine
	Subfascial
Catheter (11):	Adjacent gangrenous gall bladder (4)
	Intrahepatic (tumorabscess)
	Subhepatic (3)
	Subphrenic
	Perirenal
	Splenic bed abscess

Non cured abscesses

necrotic hollow organ (gall bladder, intestine, fallopian tube etc.). In the puncture group a double abscess in the left iliac fossa after removal of a renal transplant could not be cured by 2 punctures. At operation a subhepatic abscess showed intraperitoneally disseminated multiple abscesses. A tubo-ovarian abscess failed to heal despite 2 punctures with aspiration of 30 and 50 cc's of pus and repeated instillations of antibiotics. A retrocolic abscess was caused by a lar-

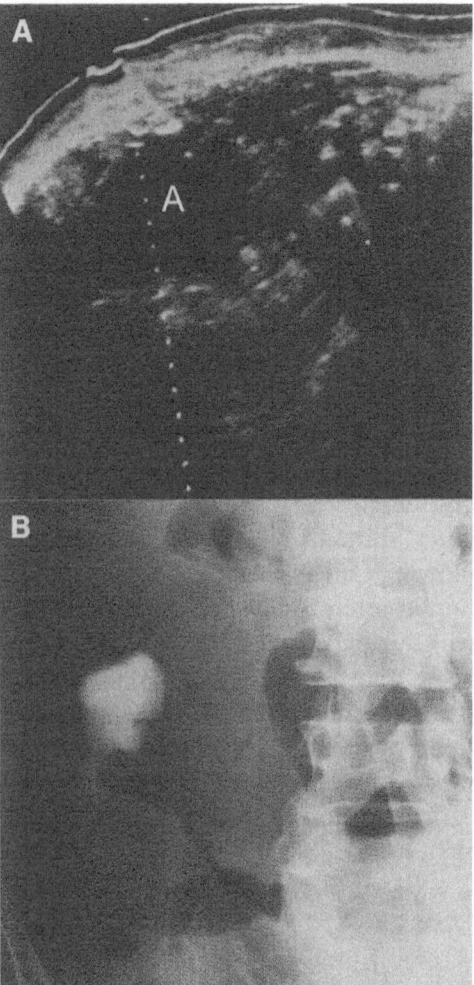

Fig. 24-4. Right iliac fossa abscess
A 69-year-old male with an abscess after appendectomy treated by 3 punctures. A: Transverse scan through the right iliac fossa. A: abscess. B: Radiograph after contrast medium injection visualizing the abscess.

ge perforation of the colon and hemicolectomy was necessary. In a subfascial abscess the patient was operated on too early after the puncture, leaving no time to allow the puncture to become curable.

In the catheter group of 4 patients with abscesses in or adjacent to a necrotic gall bladder (Fig. 24-5), the drainage

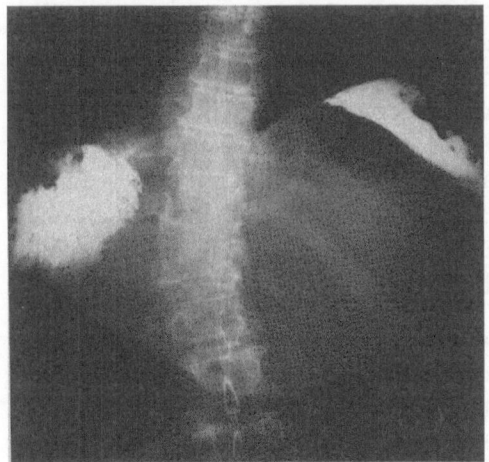

Fig. 24-3. Subphrenic abscesses
A 68-year-old female with bilateral subphrenic abscesses cured by 3 weeks of catheter drainage. Radiograph after injection of contrast medium through the catheters.

was only temporary and an elective cholecystectomy was necessary after a few days of catheter drainage. A tumor abscess in the liver caused by metastases from a seminoma was treated with some effect, but autopsy revealed a residual abscess. There were 3 subhepatic and 1 subphrenic postoperative abscesses. One of these was not cured because of malfunction of the catheter. Another had recurrence 9 days after removal of the catheter. A third had a multiloculated abscess and a fourth a pancreatic abscess. A perirenal abscess was drained only temporarily before removal of the kidney. Finally, a splenic bed abscess following splenectomy was surgically drained after 8 days of catheter treatment during which the pus showed raised contents of amylase. At operation extensive adhesions, necroses of the adjacent tissue and very thick pus was found, caused by damage to the pancreatic tail in connection with the initial splenectomy.

In 81% of the cases cultures were positive and a broad spectrum of bacteria was found. The most frequent being E. coli. 50% were aerobic (E. coli, Klebsiella, S. faecalis, Citrobacter etc.) and the other half anaerobics (Bacteroides, Streptococcus, Peptostreptococcus and Clostridium).

The abscesses contained from 5 cc's to more than 2 liters of pus, and the catheter was left in place for an average of 10 days (range: 3–28).

Five patients had shaking chills a few hours after the procedure, and one was treated because of bacteriemia (Table 24-6). No cases of peritonitis, fistulation, spreading along the catheter or bleeding were registered.

Prior to the drainage, almost all patients had recieved systemic antibiotic treatment without effect on the temperature course. In this series about 75% showed a marked fall in temperature and clinical improvement following puncture.

The conventional treatment of intraabdominal abscesses has been either surgery with large tube drainage or conservative therapy and most surgeons have previously been reluctant to accept percutaneous aspiration of abdominal abscesses. In 1953, however, McFadzean et al. used needle aspiration in the definitive treatment of hepatic abscesses. Since then a growing body of reports has been published the results of which compare favorably with those of surgical drainage.

Ultrasonically guided diagnostic puncture of abdominal fluid collections is indicated in most cases and often yields important clinical information, especially in cases of intraabdominal abscesses. Drainage of abdominal abscesses by needle aspiration or catheterization is safe and has a success rate exceeding 80%. If multicompartment abscesses are excluded, and this can be done on the basis of the scans, a succes rate close to 90% can be obtained. Therefore, this method is an advantageous alternative to conventional surgical treatment of intraabdominal abscesses.

Table 24-6.

Possible	Experienced
Shaking chills	5
Bacteremia	1
Spread along catheter	0
Contamination of pleural and peritoneal cavity	0
Bleeding	0

Complications in 108 abscesses treated by ultrasonically guided drainage.

References

Als O. Puncture of intraabdominal fluid collections. In: Holm H H, Kristensen J K, eds. Ultrasonically guided puncture technique. Copenhagen: Munksgaard, and Baltimore: University Park Press, 1980: 83.

Altemeier W A, Culbertson W R, Fullen W D, Shook C D. Intra-abdominal Abscesses. *Am J Surg* 1973; 125: 70.

Elyaderani M K, Skolnick M L, Weinstein B J. Ultrasonic detection and aspiration confirmation of intraabdominal collection of fluid. *Surg Gynecol Obstet* 1980; 149: 529.

Gerzof S G, Spira R, Robbins A H. Percutaneous abscess drainage. *Semin Roentgenol* 1981; 16: 62.

Grønvall J, Grønvall S, Hegedüs V. Ultrasound guided drainage of fluid-containing masses using angiographic catheterization technique. *AJR* 1977; 129: 997.

Grønvall S, Gammelgaard J, Haubek A, Holm H H. Drainage of abdominal abscesses guided by sonography. *AJR* 1982; 138: 527.

Haubek A, Gammelgaard J, Grønvall S, Holm H H. Ultrasonically guided percutaneous puncture and biopsy technique. In: Wilkins R A, Viamonte Jr M, eds. Interventional Radiology. Oxford: Blackwell Scientific Publications, 1982: 400.

Holm H H. Procedure of ultrasonically guided puncture. In: Holm H H, Kristensen J K, eds. Ultrasonically guided puncture technique. Copenhagen: Munksgaard, and Baltimore: University Park Press, 1980: 29.

Jensen F, Pedersen J F. The value of ultrasonic scanning in the diagnosis of intra-abdominal abscesses and hematomas. *Surg Gynecol Obstet* 1974; 139: 326.

Maklad N F, Doust B D, Baum J K. Ultrasonic diagnosis of postoperative intra-abdominal abscesses. *Radiology* 1974; 113: 417.

McFadzean A J S, Chang K P S, Wong C C. Solitary pyogenic abscess of liver treated by closed aspiration and antibiotics: a report of 14 consecutive cases with recovery. *Br J Surg* 1953; 41: 141.

Mueller P R, van Sonnenberg E, Ferrucci J T. Percutaneus drainage of 250 abdominal abscesses and fluid collection. Part II: Current procedural concepts. *Radiology* 1984; 151: 343.

Taylor K J W, Wasson J, DeGraff C, Rosenfield A T, Andriole V T. Accuracy of gray-scale ultrasound diagnosis of abdominal and pelvic abscesses in 220 patients. *Lancet* 1978; 1: 83.

van Sonnenberg E, Mueller P R, Ferrucci J T. Percutaneus drainage of 250 abdominal abscesses and fluid collection. Part I: Results, failures and complications. *Radiology* 1984; 151: 337.

CHAPTER 25

Interventional echocardiography

Ivo Cikes, Branko Breyer, Aleksander Ernst & Fedor Custovic

Interventional echocardiography can be defined as the use of echocardiography for guiding or assessing various interventions in cardiology, making them safer and less invasive.

Several interventional methods in cardiology can be guided by echocardiography: pericardiocentesis, percutaneous pericardial biopsy, percutaneous pericardial fenestration, endomyocardial biopsy, cardiac catheterization, introduction of pacemaker electrodes and electrodes for electrophysiological studies, intracardiac echocardiography, transesophageal echocardiography, transseptal left atrial catheterization, balloon septostomy, intraoperative echocardiography and laser phototherapy.

Interventions which can be assessed by echocardiography are: percutaneous transluminal coronary angioplasty, streptokinase thrombolysis, contrast studies, myocardial perfusion studies, drug studies, exercise echocardiography and intraoperative echocardiography.

Over the last few years our laboratory has been involved in developing intrapericardial interventions directed by ultrasound. Recently, a new method of cardiac catheterization and the introduction of pacemaker electrodes and

electrodes for electrophysiological studies under echocardiographic guidance have been developed.

Pericardiocentesis guided by echocardiography

Although introduced in 1840, pericardiocentesis is still a procedure with a high risk of morbidity and mortality. Before the era of echocardiography, complications were frequently reported. The introduction of echocardiography in the routine diagnosis of pericardial effusion is a most important advance in reducing the hazards of pericardiocentesis. Since echocardiography can provide a reliable diagnosis and determine the approximate amount and distribution and possible loculation of pericardial fluid, it is essential in the selection of patients for pericardiocentesis. And on this basis the optimal puncture site can be determined.

Several methods have been reported which claim to reduce the potential risks of pericardiocentesis (Fig. 25-1) (Cikes & Ernst). Fluoroscopic guidance of the puncture needle is not a reliable method because it cannot differentiate pericardial effusion from the cardiac mass. Electrocardiographic monitoring

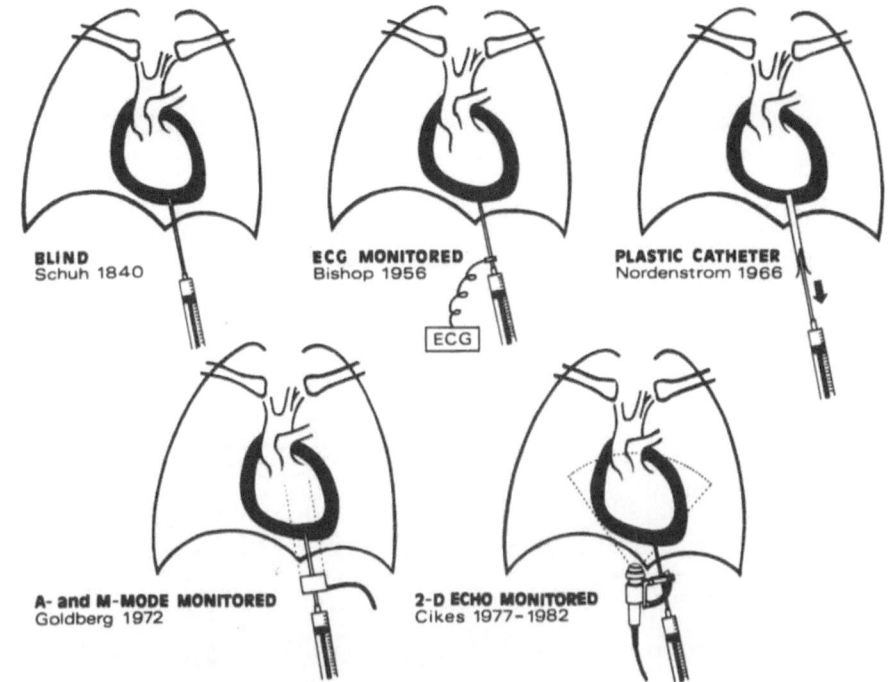

Fig. 25-1. Different techniques of pericardiocentesis

of pericardiocentesis described by Bishop and coworkers may give a spurious feeling of safety and usually registers already existing injury. Plastic catheters introduced into the pericardial sac over or through the needle or guide wire is much safer. In 1972, Goldberg & Pollock used a special transducer with a hole in the center to direct the needle during pericardiocentesis under A- and M-mode echocardiographic control. The needle tip was seen in the A- and M-mode display as an echo arising at the needle tip-fluid interface. In 1 out of 6 reported patients ventricular puncture was performed. As is seen from the literature, this method was of no further interest in cardiology. The main disadvantage of pericardiocentesis guided under A- and M-mode control is the lack of spatial orientation.

In order to ensure safe diagnostic and therapeutic pericardiocentesis we have, over the last 5 years, guided the needle for pericardiocentesis by 2-dimensional echocardiography in 25 patients with pericardial effusion. All the patients had moderate to large pericardial effusions. For safe pericardiocentesis it is essential to position the needle in the scanning plane. If part of the needle is outside the scanning plane the location of the needle tip is misinterpreted (Fig. 25-2). The position of the needle tip can be confirmed by a contrast study – instillation of 2 to 5 ml of sterile saline or pericardial fluid through the needle. If a contrast jet appears at the location of the presumed needle tip, it can be considered the true tip (Fig. 25-3). Besides the classical subxyphoid approach we perform pericardiocentesis through the

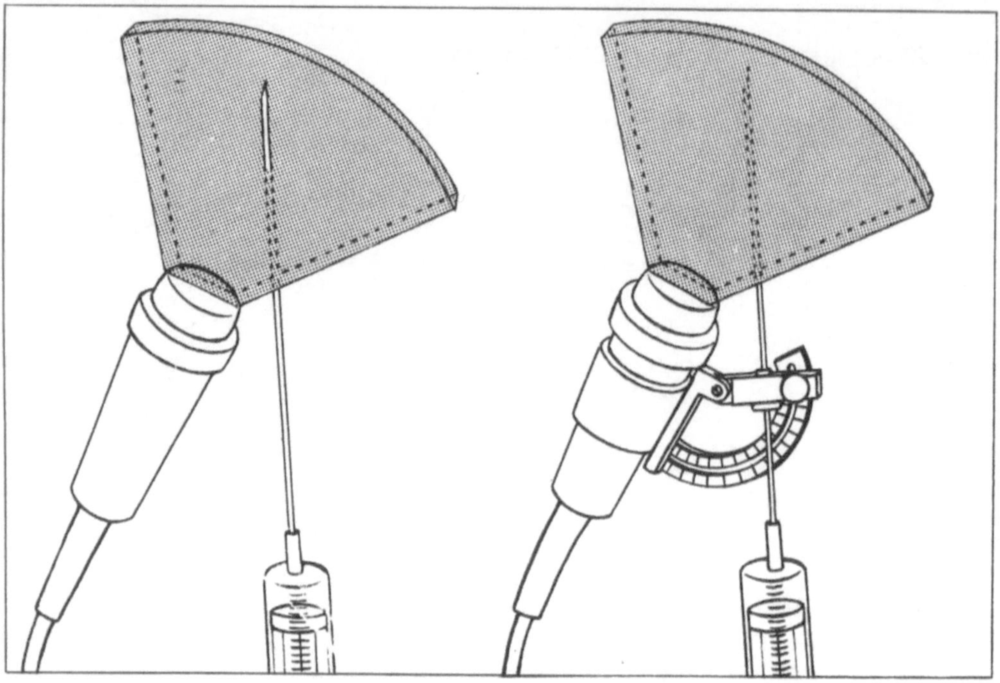

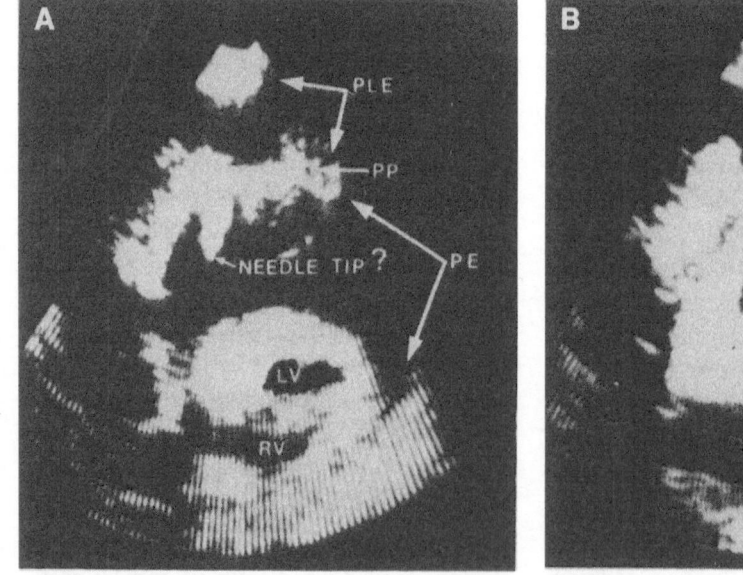

Fig. 25-2. Identification of catheter tip
The point where the needle leaves the scanning plane may be misinterpreted as the needle tip. A puncture adaptor (right) mounted on the transducer ensures that the needle is kept in the scanning plane at an adjustable angle to the transducer.

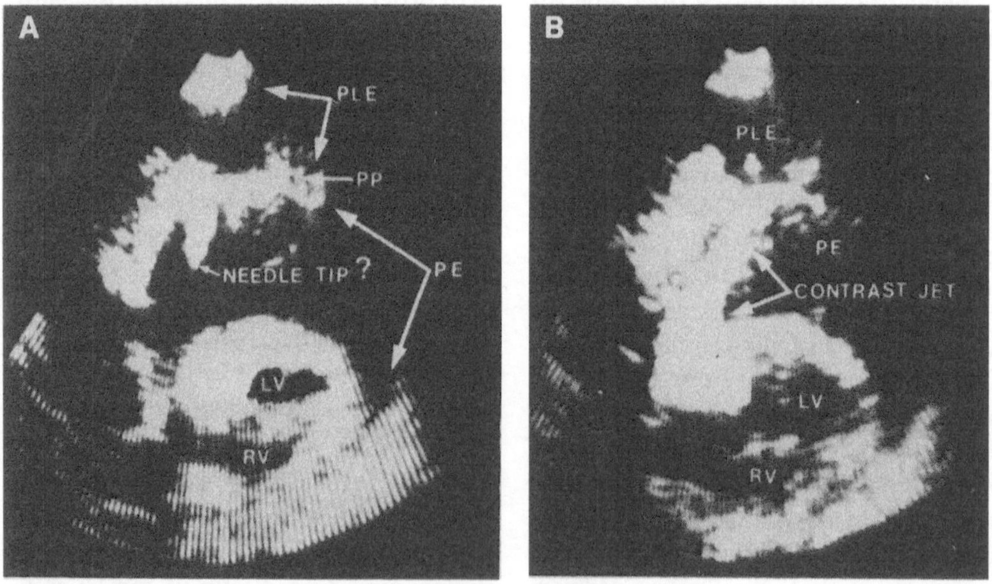

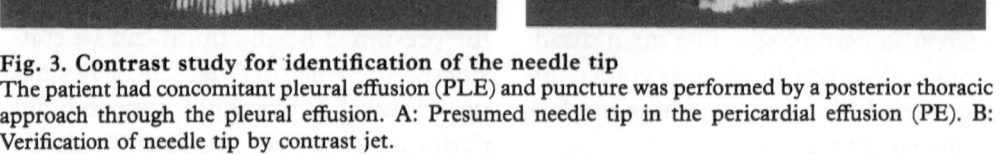

Fig. 3. Contrast study for identification of the needle tip
The patient had concomitant pleural effusion (PLE) and puncture was performed by a posterior thoracic approach through the pleural effusion. A: Presumed needle tip in the pericardial effusion (PE). B: Verification of needle tip by contrast jet.

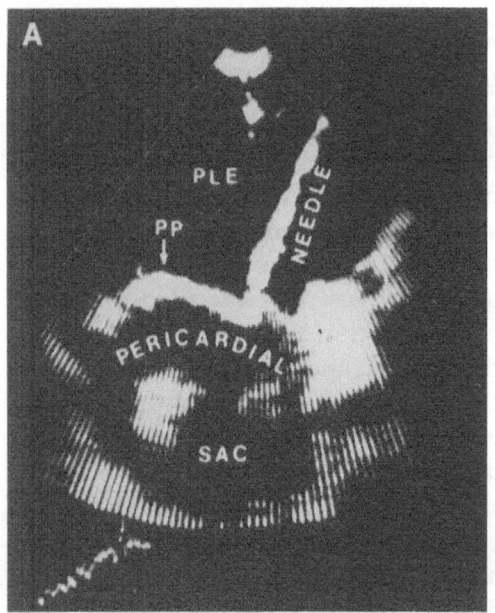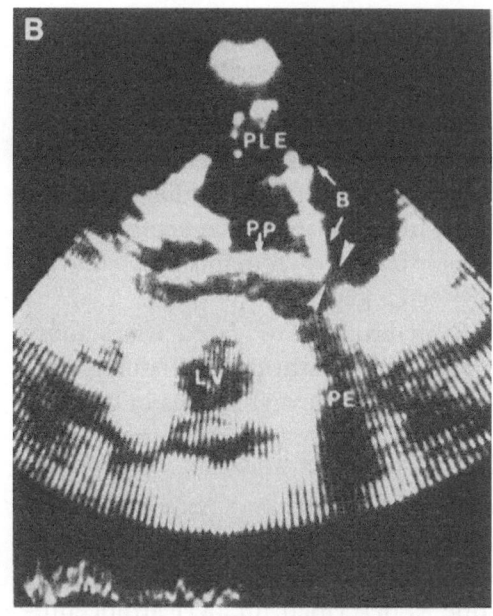

Fig. 25-4. Pericardial window
A: The needle for pericardiocentesis in pleural effusion (PLE) reaching parietal pericardium (PP). B. Bioptom (B) and window (arrows) in the parietal pericardium created during pericardial biopsy. PE: pericardial effusionl LV, left ventricle.

posterior or lateral thoracic wall in patients with concomitant pleural effusion (Fig. 25-4). We find this approach to be most convenient provided a large pleural effusion assures a large echo-free corridor to the parietal pericardium. In addition, the largest amount of pericardial fluid usually collects behind the left posterior ventricular wall and the cardiac apex moves away from the needle in systole, reducing the risk of cardiac damage. After entering the pericardial sac the needle should be replaced by a plastic catheter. The position of the needle or replacement catheter should be monitored continuously during the procedure and the needle repositioned if necessary. In 1 out of our 25 patients a ventricular puncture with intrapericardial bleeding occurred, requiring im-

mediate thoracotomy. In this patient an apical approach was used because an enlarged nodular liver thwarted the subxyphoid approach.

Recently, we have constructed a puncture adaptor mounted on the phased array transducer (see Fig. 25-2). The adaptor ensures that the needle follows the predetermined needle path and stays in the scanning plane. This eliminates the possibility of misinterpretation of the position of the needle tip. The angle of the needle path can be adjusted. Particular attention should be paid to preparation of the patient, adequate sterilization of instruments and electrical safety precautions. We believe that the technique described can eliminate the risk of cardiac damage during pericardiocentesis.

Percutaneous pericardial biopsy and fenestration

Diagnostic pericardiocentesis enables etiologic diagnosis in less than one third of patients with pericardial effusion. Thus, in many cases thoracotomy with pericardial biopsy remains the definitive diagnostic solution.

In an attempt to avoid thoractomy and general anesthesia for histologic diagnosis of pericardial lesions in patients with pericardial effusion we have introduced a new technique of percutaneous pericardial biopsy and pericardial fenestration under 2-dimensional echocardiographic guidance. The same technique as described for pericardiocentesis was used. Pericardial biopsy was performed in 6 patients, while a pericardial window was created in 2 patients. In patients with concomitant left pleural effusion, the posterior or lateral thoracic approach was preferred, as in pericardiocentesis (Fig. 25-4). A disposable Trucut Travenel biopsy needle with a 15.2 cm cannula length and 20 mm specimen notch was used. After the pericardial specimen was taken the hole was enlarged to a pericardial window and pericardial fluid drained into the pleural space. A pericardial window created during biopsy for decompression can save repeated pericardiocentesis in recurrent pericardial effusion with tamponade. In all patients an adequate tissue specimen for histological analysis was obtained and proper histological diagnosis was made. It is believed that the adaptor described will refine the biopsy technique and improve its safety.

Cardiac catheterization guided by ultrasound

To avoid the well-known disadvantages of conventional X-ray guided cardiac catheterization, we have developed a system for ultrasonic guidance of the catheters in cardiac cavities. The main disadvantages of X-ray guided cardiac catheterization are: X-ray irradiation of the patient and laboratory staff, injection of a large volume of iodine contrast with its toxic and hemodynamic effects, lack of spatial orientation, lack of detailed anatomical data, the high cost of equipment, space and installation. This method can therefore be used only in a limited number of cardiac centers.

Ultrasonically guided cardiac catheterization can overcome all these disadvantages. The introduction of catheters into cardiac cavities is not, in itself, the most invasive part of cardiac catheterization, as arrhythmias can now be kept well under control. Two-dimensional echocardiography provides superior morphologic information of the heart in real-time which could not be obtained by other techniques. However, in some patients we need additional data on intracardiac pressures and oxymetric data for the calculation of intracardiac shunts in order to make a definitive cardiac diagnosis and clinical decisions.

The portions of the catheter which are in the scanning plane can be ultrasonically imaged in the cardiac cavities due to their different acoustic impedance. In order to obtain intracardiac pressure measurements and blood samples for a precise position it is essential to identify the catheter tip (Fig. 25-5). There are several possible sources of misinterpretation of the catheter tip.

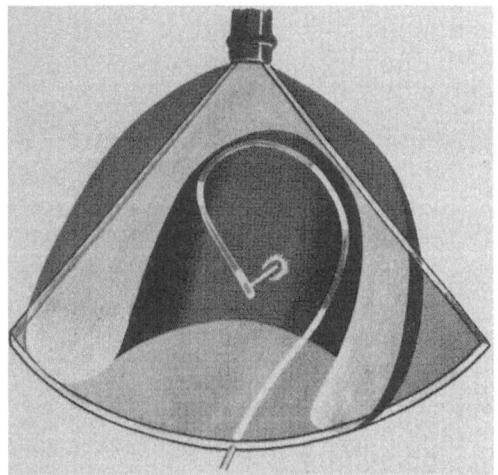

Fig. 25-5. Catheter tip misinterpretation
Each point where the catheter is leaving or entering the scanning plane can be misinterpreted as the catheter tip.

Each point at which the catheter enters or leaves the scanning plane can be misinterpreted as the tip. Some cardiac structures such as the papillary muscles, trabeculae, chordae tendineae and the valvular apparatus can also be sources of misinterpretation of the catheter tip. A possible problem may also be phantom or spurious echoes.

We have tested several method for the identification of the catheter tip. Injection of contrast through the catheter has proved to be an unreliable method as the catheter tip during injection usually moves out of the scanning plane whereby part of the catheter can be misinterpreted as the tip. Marking of the catheter tip with a material of different acoustic properties acting as a passive reflector was promising in *in vitro* experiments, but unsatisfactory *in vivo*. This is because of the numerous echoes from the already-mentioned cardiac structures or spurious echoes which obscure the image of the tip reflector.

We have solved the problem of identification of the catheter tip by mounting a cylindrical miniature transducer unit and by the use of associate electronics. The catheter transducer can act in an active or passive mode. In the active system an external pulse generator connected to the catheter transducer induces ultrasonic pulses which are picked up by the scanner transducer. These appear on the 2-dimensional echocardiographic image as a bright line always crossing the catheter transducer (Figs. 25-6, 25-7). The main advantages of this system are simplicity and independence of the scanner. Its disadvantages are the introduction of relatively high voltages (up to 10 V) into the heart (necessitating special safety insulation) and the lack of depth resolution.

In the passive system the catheter transducer acts as a receiver of the ultrasound pulses from the external transducer. The principle of operation is illustrated in Fig. 25-8. The catheter transducer is triggered by ultrasound pulses from the scanner transducer and the voltage generated is transmitted to the timing electronics where it can be amplified and used for timing. The time between ultrasound transmission from the scanner transducer and detection by the catheter transducer must be doubled in order to position the marker signal correctly. As a result a wel-defined blinking marker adjacent to the catheter tip can be seen on the 2-dimensional echocardiographic image (Figs. 25-9 & 25-10). The blinking marker can be easily differentiated from the cardiac structures and is visible at any gain setting. The passive system has the advantages that it does not introduce electrical pulses into the heart, it has the best marker

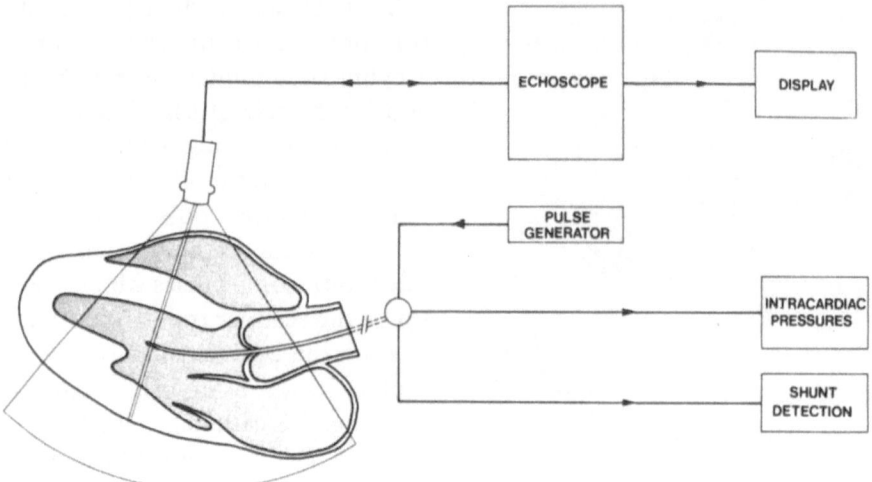

Fig. 25-6. Active system for marking of the catheter tip
A transducer at the tip of the catheter emits ultrasonic pulses which are picked up by the scanner transducer producing a bright line on the image (Fig. 25-7) corresponding to the catheter tip.

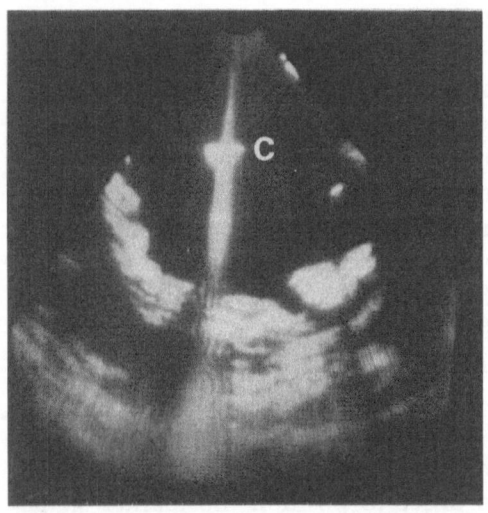

Fig. 25-7. Orthogonally sected catheter (C) in water-bath with marking line

properties and a good axial and lateral resolution. Its only disadvantage is scanner dependence such that the timing electronics input and output must be differently designed for different scanners.

The efficacy of the prototype system was tested in a water-bath, in cadaveric hearts and in beating dog hearts. In all instances the blinking marker of the catheter tip was easily identified and correctly positioned. With this system it seems that echocardiography can be used not only for obtaining unique morphologic data but also for additional data on intracardiac pressures and oxymetric data for the calculation of intracardiac shunts. Angiography may be performed using indifferent echo-contrasts. The same system can be used in guiding and assessing myocardial perfusion studies.

The advantages of echo guided cardiac catheterization over conventional X-ray guided catheterization are the avoidance of ionising radiation, as well as of the toxic and hemodynamic effects of iodine contrast medium, better spatial anatomy due to the multiple sectional planes which allow for precise positioning of the catheter tip, and finally the fact that this method is considerably less

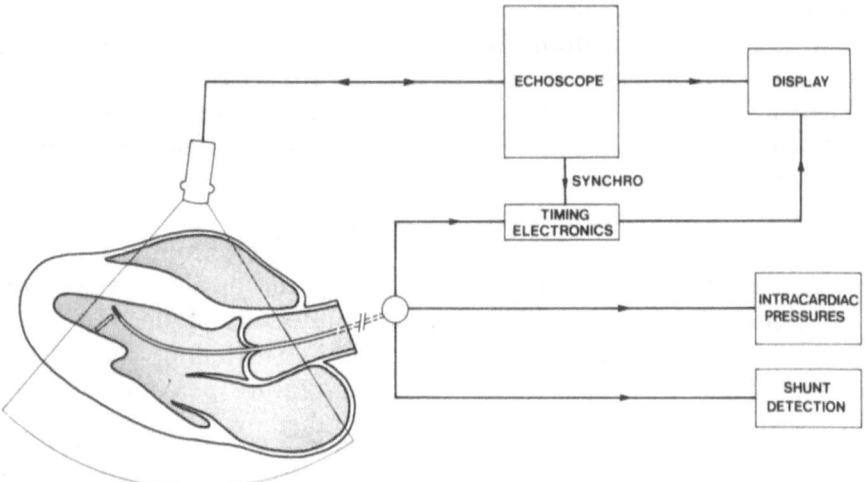

Fig. 25-8. Passive system for marking of the catheter
The catheter transducer acts as a reciever of the ultrasound pulses from the external transducer.
Thereby the catheter tip is indicated by a short line (S in fig. 25-9).

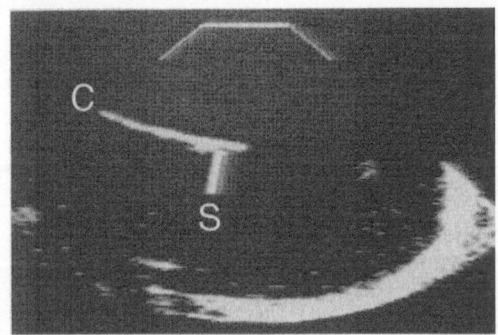

Fig. 25-9. The catheter (C) in water-bath with the signal (S) of the passive marking system

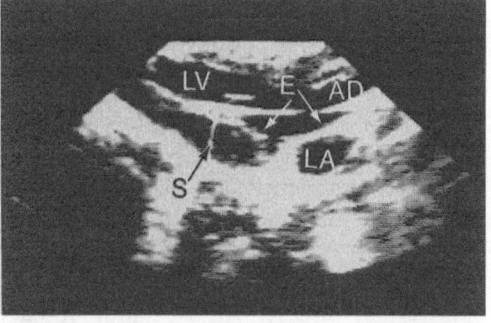

Fig. 25-10. Long-axis section of beating dog heart
Catheter (C) and signal (S) of the passive marking system. LV: left ventricle, Ao: aorta, LA: left atrium.

expensive, allowing its use in smaller centers. Disadvantages are lack of coronary angiography and poor images in some patients with a small "echo-window".

It seems likely that after a clinical trial the method can become a procedure for complete cardiac diagnosis; in the majority of patients it can possibly replace X-ray guided cardiac catheterization.

Ultrasonically guided introduction of pacing and electrophysiologic electrodes

Recently, we have used the passive system for marking the tip of pacemaker and electrophysiologic electrodes. It allows electrode introduction and precise positioning under echo guidance. It may also be used for better diagnosis of acute and chronic pacing system malfunction. Precise diagnosis of dislodgement, microdislodgement and myocardial perforation is possible. The M-mode technique has been found to be superior for the identification of electrode microdislodgement.

A pacemaker has been adapted for telemetry of pulses from the electrode transducer, allowing for additional electronic control. Thus, non-invasive discrimination between lead fracture (X-ray invisible) and pulse generator failure in cases of loss of pacing pulse is possible. As high frequency control of the echo signal is more sensitive on lead impedance alternation, the insulation break detection in early phase is possible.

Marking electrophysiologic electrodes by the passive system provides exact placement of the His electrode, endocardial mapping, and perhaps in the near future percutaneous catheter ablative therapy.

References

Cikes I. Interventional echocardiography. 5th Symposium on Echocardiology. Rotterdam, 1983: Abstracts p. 38.

Cikes I, Ernst A, Breyer B. Interventional echocardiography, 3rd International Congress on Interventional Ultrasound. Copenhagen, 1983: Abstracts p. 102.

Cikes I, Ernst A. New aspects of echocardiography for the diagnosis and treatment of pericardial disease. In: Roelandt J, ed. The practice of M-mode and two-dimensional echocardiography. The Hague, Boston, London: Martinus Nijhoff Publishers, 1983: 141.

Cikes I, Breyer B, Ernst A, Custovic F. Cardiac catheterization guided by ultrasound, *JACC* 1984; 3: 564.

Cikes I. Echocardiography to guide interventions, 4th International Congress on Echocardiology, Verona, 1984; Abstracts p. 35.

Hanrath P, Bleifeld W, Souquet J, eds. Cardiovascular diagnosis by ultrasound. Transesophageal, computerized, contrast, Doppler echocardiography. The Hague, Boston, London: Martinus Nijhoff Publishers, 1982.

Cikes I. New echocardiographic possibilities in the etiological diagnosis and therapy of pericardial diseases. In: Hanrath P, Bleifeld W, Souquet J, eds. Cardiovascular diagnosis by ultrasound. The Hague, Boston, London: Martinus Nijhoff Publishers, 1982: 188.

Roelandt J, ed. The practice of M-mode and two-dimensional echocardiography. The Hague, Boston, London: Martinus Nijhoff Publishers, 1983.

Cikes I, Breyer B. Complete cardiac catheterisation guided by ultrasound. *Eur Heart J* 1983; 4 (suppl. E): 21.

Mortensen S A, Egeblad H. Endomyocardial biopsy guided by cross-sectional echocardiography. *Br Heart J* 1983; 50: 246.

Breyer B, Cikes I. Ultrasonically marked catheter – a method for positive echographic catheter position identification. *Med Biol Eng Comput* 1984; 22: 268.

Fine-needle aspiration biopsy: are there any risks?

Edward H. Smith

The field of diagnostic radiology has recently witnessed an extraordinary technologic explosion with ultrasound, computerized tomography, digital vascular imaging and nuclear magnetic resonance added to the diagnostic medical armamentarium. Despite these major advances the need for tissue verification of a suspected lesion is still required before definitive, often radical, therapy is begun. Because of its simplicity, speed, low cost and, above all, its apparent safety, fine-needle aspiration biopsy (FNABx) has increasingly become the method of choice to establish the cytologic diagnosis of a pathologic process in the abdomen and elsewhere in the body.

This chapter will review the evidence available in order to determine the relative safety of fine-needle aspiration biopsy. In addition to a careful review of the literature, a survey was conducted, the results of which will be reported here.

Experimental evidence of tumor spread after fine-needle aspiration biopsy – literature review

Engzell and coworkers conducted a series of experiments where $V \times 2$ carcinoma was transplanted to the tarsal area of rabbits following which popliteal node spread occurred. The standard aspiration technique using an 18-gauge needle was employed, involving suction and vertical agitation. Following biopsy, the lymph node was massaged and the fluid expressed was collected and examined along with the lymphatic and venous drainage of the involved node. Smear of the fluid was positive for malignancy in 10 of 12 animals.

A second study was conducted by Struve-Christensen who, immediately after pneumonectomy or lobectomy for tumor, injected dye up to the margin of the tumor of the excised lung and then biopsied the lesion with a 19-gauge (1 mm) needle. He then carefully excised the needle tract right down to the tumor mass and examined the sections histologically. His material included 27 cases where malignant lesions were aspirated and, in 24 of these (89%), malignant cells were demonstrable in the needle tract.

In a rather ingenious study Ryd and his colleagues performed FNABx of solid tumors in mice with the 23-gauge needle first passing through the leg muscle of a healthy animal into the tumor of the donor animal and then withdrawn through both. Needle tract seed-ing occurred in 65–83% of the trans-muscularly punctured animals. Similar experiments were carried out with as-cites tumors with the needle passing through the abdomen of the recipient mouse into the abdomen of the donor (tumor-bearing) mouse and then with-drawn through both with a 90% "take". (A semiquantitative method was also performed labelling the tumor cells with a radioactive label and performing fine-needle aspiration of the malignant asci-tic fluid through the leg of a recipient mouse). However, as the authors point out, this experimental model is not analogus to the clinical situation in hu-mans, since this particular animal tumor was highly malignant, contained very little stroma and lacked intercellular junctions, allowing large numbers of cells to be easily detached by the needle.

It is apparent from these experimen-tal studies that tumor cells probably do leak out into the tissues and are de-posited along the needle tract after as-piration biopsy in a large percentage of cases. However, in view of the (appar-ently) rare clinical instances of needle tract seeding, it has been assumed that these tumor cells, in the great majority of cases, are destroyed by the host im-mune response, or by some other mech-anism.

Other complications – animal experimentation

To determine the incidence of hemato-ma and bowel perforation following as-piration biopsies, Goldstein et al. per-formed multiple percutaneous aspi-rations of abdominal organs in dogs. At exploration no evidence of significant hematoma formation or bowel perfo-ration was noted.

Clinical evidence of hazards – literature review

A recently published review of the literature detailed the risks of FNABx in the abdomen. The authors were able to tabulate the results of 11,000 FNABx including their own 552 patients and discovered 1 death, 2 instances of needle tract seeding (these 3 cases are included in the present literature search), and 4 other "major" complications, as well as 58 relatively minor ones. They calcu-lated the mortality rate of FNABx to be 0.008%, and the incidence of other major complications to be 0.05% (Liv-raghi et al.).

A 10-year follow-up of 157 patients who had undergone fine-needle aspi-ration biopsy of salivary gland aden-omas using a 22-gauge needle was car-ried out by Engzell. Follow-up included clinical examination and autopsy results in those patients who did not survive. No evidence of needle tract seeding was discovered. Another study by Engzell and colleagues included a 5-year follow-up of 469 patients undergoing transrec-tal biopsy of prostatic carcinoma using a 22-gauge needle. One instance of local rectal recurrence was noted 3 years after biopsy and was thought unlikely to be secondary to needle tract seeding.

Engzell also quotes a study by Esposti which consisted of a 5-year follow-up of 656 patients who had undergone fine-needle aspiration biopsy of cervical lymph metastases with no evidence of seeding.

Many large series have been reported with much shorter follow-up without evidence of significant complications, including needle tract seeding (Kline & Neal, Ho et al., Lalli et al., Lundquist, Holm et al.).

Von Schreeb and coworkers compared the 5-year survival of 2 groups of approximately 75 patients each with renal cell carcinoma; the groups differed only in that one underwent aspiration biopsy and injection of radiographic contrast. The 5-year survival in the punctured group was actually considerably better than in the non-punctured group. Certainly no adverse effects of aspiration biopsy could be detected.

A 15-year actuarial survival rate was calculated for 370 patients operated upon for breast cancer following aspiration biopsy and this was compared to a closely matched control group. The survival rates were identical (Berg & Robbins).

Needle tract seeding – literature review

Only 2 cases of needle tract seeding after thin-needle aspiration biopsy have been reported. There are numerous reports of needle tract seeding and other complications following large-bore cutting needle biopsy, but this review will deal only with fine-needle aspiration which conventionally includes needles of 20-gauge and smaller.

A single case of needle tract seeding was reported in 1979 by Ferrucci and colleagues who described a patient with pancreatic carcinoma who underwent needle aspiration biopsy under CT guidance using a 22-gauge needle. The lesion was difficult to biopsy and 10 passes were required. Three months later, a skin nodule was noted at the needle insertion site which was confirmed by biopsy.

The 2nd instance of needle tract seeding followed 22-gauge needle aspiration biopsy in a patient with a pancreatic carcinoma, as reported by Smith and colleagues who also noted a 3-month interval between the biopsy and the needle tract nodule.

Needle tract seeding occurred 2 years after aspiration biopsy of a renal cell carcinoma. However, in this case an 18-gauge spinal needle was used which does not come under our definition of "fine-needle" (Bush et al.).

Fatalities – literature review

Two fatalities have been reported. One was that of a patient who died after fine-needle aspiration of the liver using a 0.7 mm (22-gauge) needle. The patient, who had cirrhosis and a hepatoma (verified by the needle aspiration), hemorrhaged following the biopsy and died 2 days later (Riska & Friman).

A 2nd death occurred following aspiration biopsy of the pancreas using a 22-gauge needle in a patient who developed necrotizing pancreatitis following biopsy. Death occurred 5½ months after the biopsy and at autopsy the pancreas was seen to be free of tumor. The authors speculate that it may be more hazardous to biopsy a normal pancreas which could leak enzymes than a pancreatic adenocarcinoma which does not form enzymes (Evans et al.).

Other serious complications have been reported including hemorrhage, gram-negative sepsis, acute peritonitis and shock, bile leak with peritonitis, perforated stress ulcer, pneumothorax, and broken needles (Ferrucci et al.,

Schnyder et al., Livraghi et al., Schulz, Otto, Silver & Thornbury). Vasovagal reactions simulating hemorrhage have occurred, leading to unnecessary surgical exploration and death. Hypotension with bradycardia or without tachycardia should suggest this mechanism (Bigongiari).

Questionnaire

Because of the sparsity of reports in the literature concerning complications of fine-needle aspiration biopsy, it was decided to survey hospitals by means of a questionnaire in an attempt to determine whether the widely held opinion of the innocuousness of fine-needle aspiration biopsy was justified.

A questionnaire was sent (Fig. 26-1) to over 100 university hospitals in the United States and, in addition, to every 5th hospital (over 200 beds) from a list of hospitals in the United States (supplied by a commercial vendor of X-ray film). Eight hospitals outside the United States were also sent questionnaires because of their reported experience with the procedure. Preliminary analysis of the questionnaire results will now be reported (as the data are still incomplete). Of the 479 questionnaires sent, 255 responded. Of these, 41 reported that they did not perform the procedure, leaving 214 (45%) for analysis.

The 214 responders reported at total of 15,777 aspiration biopsies carried out for the year in question. Analysis of a sample of the responders (57) revealed that the procedure had been employed for an average of 4 years. Using this figure, a total of 63,108 fine-needle aspiration biopsies was assumed.

Results – complications listed in Tables 26-1–26-4.

As best as can be determined these complications involve the use of fine-needle (20–23 gauge) and most often 22 or 23 gauge and do not include "cutting" types of needles such as Menghini or Trucut. Unquestionably, some of the latter have been inadvertently included as in some instances the type of needle used was either not indicated or was ambiguously listed.

Hemorrhage

There were 27 cases of clinically significant hemorrhage, i.e. requiring blood transfusion. Three of these were fatal, and will be discussed in detail subsequently. Bile leaks occurred in 51 patients. In all likelihood, this figure is inflated by the inclusion of transhepatic cholangiography; these were excluded when specifically mentioned by responders. Four instances of bile peritonitis were reported, but this figure is undoubtedly low since this complication was not specifically sought in the questionnaire.

The series included 16 generalized infections, none of which were fatal.

Needle tract seeding

There were 3 occurrences of needle tract seeding, including 1 which has already been published by Ferrucci (Table 26-1). In the latter case, as has been mentioned earlier, a pancreatic carcinoma was biopsied under CT guidance. The case was unusual in that 10 needle passes were required, on 2 separate occasions, before malignant cells could be

QUESTIONNAIRE

DIRECTIONS: Please circle appropriately or complete answer as otherwise indicated.

1. Type of hospital: University   Proprietary   Veterans   Non-Profit   Other _____

2. Number of beds : ·   200+   300+   400+   500+   ·

3. Number of individuals performing aspiration biopsies: _____

4. Type of guidance: (a) Blind   (b) CT   (c) US   (d) Fluoro   (e) Other _____
        Relative percentage: _____   (specify)

5. If ultrasound: (a) Real Time (b) Static (c) Biopsy Transducer   Static _____
                                                                        Real Time _____

6. Number of aspiration biopsies carried out per year: _____
    Number of years performing biopsies: _____

7. Size of needle most often used: Gauge: #18   #19   #20   #21   #22   #23   Other_____
    If more than one needle used, please indicate relative percentages:
            _____ %   _____ %   _____ %

8. Type of (non-cutting) needle(s) used: (a) Spinal   (b) Chiba   (c) Other _____

9. Average number of passes: 1   2   3   4   5   more____

10. Types of lesions aspirated:
        A. Neoplasms of: (a) pancreas (b) liver (c) kidney (d) adrenal (e) spleen
                         (f) lymph nodes   (e) other _____

        B. Abscesses. :

11. COMPLICATIONS ENCOUNTERED: (please specify number and describe briefly, including
                               needle size)

        A.____Needle tract seeding.

        B.____Dissemination of malignancy within 3 months (if so, time interval, type
              of tumor aspirated, size of needle used, and number of passes).

        C.____Hemorrhage - clinically significant, i.e. requiring blood transfusions,
              or surgical intervention.

        D.____Infection - clinically significant, Local:

        E.____Infection - clinically significant, Generalized:

        F.____Other

        G.____Bile Leak

        H.____Death

12. Specifically, have you biopsied a transitional cell carcinoma of the kidney or
    bladder?   YES   NO   Number_____

13. If Yes, any seeding or dissemination?   YES   NO   Number_____

14. Have you aspirated a case of: Tuberculosis   Histoplasmosis   Coccidioidomycosis
                                                YES   NO   Number_____

15. Have you aspirated a parasitic abscess?   YES   NO   Number_____

16. Any evidence of dissemination?   YES   NO   Number_____

17. Have you aspirated ovarian tumors?   YES   NO   Number_____

18. Any evidence of dissemination?   YES   NO   Number_____

Fig. 26-1.

Table 26-1. Needle tract seeding

Lesion Bx'd	Needle Size	Type of Guidance	Number of Passes	Time Interval	Comments
1. Pancreatic CA	#22	CT	10 (2 days)	3 mos.	Previously published
2. CA of the cervix	#22	Palpation	3	3 mos.	
3. Renal CA	#20	CT	?	?	Needle tract seeding noted at surgery.

recovered. A 22-gauge needle was used, and the needle tract seeding occurred 3 months following the biopsy. The second case occurred in a patient with a palpable pelvic mass extending into the lower abdomen. Percutaneous biopsy was performed using a 22-gauge Chiba needle. Three needle passes were made without the use of external guidance. Three months later the tumor extended to the needle site, breaking through the skin. Tissue diagnosis was compatible with carcinoma of the cervix. The third instance was less well documented, and involved aspiration biopsy of a renal carcinoma with a 20-gauge needle. The radiologist who performed the biopsy was told that needle tract seeding was noted at surgical exploration, but the time interval was not known. The biopsy was performed under CT guidance, but the number of needle passes was not recorded.

The 3 cases of needle tract seeding occurred in 63,108 cases for an inci-

dence of 0.005% or approximately 120,000.

Fatalities

A total of 4 deaths was reported among the 63,108 cases for an incidence of 0.006% or approximately 1:115,000 (Table 26-2).

The first death occurred following biopsy of an adrenal angiosarcoma under ultrasound guidance using a 22-gauge needle. Hemorrhage occurred following biopsy, and although the patient was transfused 8 units of blood and underwent surgical exploration, hemorrhage continued and the patient expired. The 2nd fatality occurred in a patient with multiple myeloma who, despite receiving a platelet transfusion to correct a low platelet count, hemorrhaged to death. The 3rd was in a patient with small cell carcinoma of the lung with liver metastases. The liver lesion was biopsied with a 22-gauge need-

Table 26-2. Deaths

Lesion Bx'd	Needle Size	Type of Guidance	Mode of Death	Comments
1. Adrenal angiosarcoma	#22	US	Hemorrhage	Transfused 8 units blood, died despite surgery.
2. Myeloma	#22	US	Hermorrhage	Low platelets, corrected by platelet transfusion prior to biopsy.
3. Liver met. from lung 1°	#22	US	Hemorrhage	Family refused therapy.
4. Pancreas	#22	CT	Pancreatitis	Colon-pancreas fistula transected during subsequent surgery (see text)

le under ultrasound guidance. Hemorrhage occurred and the patient's family refused further therapy because of the patient's terminal condition. The 4th death could only indirectly be attributed to the fine-needle aspiration biopsy. The patient was thought to have an enlarged pancreatic head and biopsy was undertaken using a 22-gauge needle under CT guidance. The cytologic aspirate was free of malignant cells and the immediate post-biopsy course was uneventful. Approximately 6 weeks following the biopsy, the patient underwent gastric resection for an ulcer when a "tract" between the colon and pancreas at the site of the needle biopsy was discovered and subsequently transected. Within hours of the surgery, fulminating hemorrhagic pancreatitis occurred with numerous complications including sepsis, and the patient died approximately 6 weeks later. Autopsy confirmed the tract between the colon and pancreas at the needle biopsy site. No tumor was present.

Table 26-3. Hazards of fine-needle aspiration biopsy: miscellaneous

Lesions Biopsied	#	Results
Transitional CA		
Bladder/kidney	105	No complications
Ovarian tumor*	199	No complications
Parasitic abscess	166	No complications

*mostly metastatic lesions

Miscellaneous

Some clinicians have refrained from biopsying transitional cell carcinoma of the genitourinary tract because of the fear of needle tract seeding and dissemination. Reports of these complications have not been found nor have they been registered in the 105 lesions biopsied as reported in the questionnaire (Table 26-3).

Although no complications were registered following aspiration biopsy of ovarian tumors, most of those biopsied appear to have been metastatic deposits or solid lesions. No conclusions can be drawn as to the safety of aspirating primary cystic ovarian neoplasms (Table 26-3).

Similarly, 166 cases of parasitic abscesses were aspirated without complication, including 11 cases of hydatid cysts, the latter often being regarded as an absolute contraindication to biopsy (Table 26-3). Again, no conclusions can be drawn without more data.

Overall Results

Although the overall complication rate is low (16:10,000), the 4 deaths and 3 instances of needle tract seeding are surprising and somewhat unexpected, judging from the several large series in the literature (Table 26-4).

Table 26-4. Complications of fine-needle aspiration biopsy: overall survey results

	Number	Incidence	(Approx)
Needle tract seeding	3*	0.5:10,000	(1:21,000)
Death	4	0.6:10,000	(1:16,000)
Hemorrhage	27**	4.0:10,000	(1:2,300)
Bile leak	51+	8.0:10,000	(1:1,250)
Infection	16	2.5:10,000	(1:4,000)
Total	101	16:10,000	(1:625)

  * Includes 1 previously published case.
** Includes 3 deaths.
   Probably inflated by cases of transhepatic cholangiography.

Discussion

The actual complication rate is probably higher than recorded here since most responders to the questionnaire have in all likelihood estimated their complications, rather than having systematically reviewed their patient records and might easily have missed instances of needle tract seeding or a delayed fatality because of the time interval involved. Undoubtedly, the complication rate is considerably less than for larger bore cutting-type biopsy needles (Zamcheck & Klausenstock) or for exploratory laparotomy, but it is apparent that the procedure is not entirely innocuous and should not be regarded as such. The question arises as to how to reduce this low, but not insignificant complication rate. It is logical to assume that the number of complications should be related to the number of needle passes made during the biopsy procedure. It is customary for most people to make between 3 and 5 passes to ensure adequate sampling so that the procedure need not be repeated. However, it has been recently shown by Kidd et al. that with the use of a commercially available rapid stain technique, cytologic diagnosis can be made minutes after the biopsy. If malignant cells are present in the initial aspirate, obviously repeat needle passes are unnecessary. Further experience with this technique is required before it is universally adopted. It is apparent that there is a trend towards using larger needles and small cutting-type biopsy needles which produce a small core rather than a fluid aspirate. It is important to determine the relative safety of these needles before abandoning the use of the small gauge non-cutting needles which have been proven to be quite satisfactory.

Finally, although needle tract seeding is still a rare complication, it has been suggested by Ryd et al. that the needle tract be excised when surgical removal of the primary tumor is carried out, especially when the cell type is considered highly malignant.

Conclusion

Review of the literature and the results of the hospital survey substantiate the widely held belief that fine-needle aspiration biopsy is a very safe procedure. However, serious and even fatal complications, although rare, can and do occur and it is important to be aware of the possibility and to take all the appropriate precautions in order to reduce their incidence.

References

Engzell U, Esposti P L, Rubio C, Sigurdson O, Sajicek J. Investigation on tumor spread in connection with aspiration biopsy. *Acta Radiol Ther Phys Biol* 1971; 10: 385.

Struve-Christensen E. Iatrogenic dissemination of tumor cells. *Dan Med Bull* 1978; 25: 82.

Ryd W, Hagmar B, Eriksson O. Local tumor cell seeding by fine-needle aspiration biopsy: A semiquantitative study. *Acta Pathol Microbiol Immunol Scand* 1983; (A) 91: 17.

Goldstein H M, Zornoza J, Wallace S, et al. Percutaneous fine needle aspiration biopsy of pancreatic and other abdominal masses. *Radiology* 1977; 123: 319.

Livraghi T, Damascelli B, Lombardi C, Spagnoli I. Risk in fine-needle abdominal biopsy. *J Clin Ultrasound* 1983; 11: 77.

Esposti P L. Cytologic malignancy grading of prostatic carcinoma by transrectal aspiration biopsy. A five-year follow-up study of 469 hormone-treated patients. *Scand J Urol Nephrol* 1971; 5: 199.

Kline T S, Neal H S. Needle aspiration biopsy: A critical appraisal. Eight years and 3,267 specimens later. *JAMA* 1978; 239: 36.

Ho C S, McLoughlin M J, McHattie J D, Tao L C. Percutaneous fine needle aspiration biopsy of the pancreas following endoscopic retrograde cholangiopancreatography. *Radiology* 1977; 125: 351.

Lalli A F, McCormack L J, Zelch M, et al. Aspiration biopsies of chest lesions. *Radiology* 1978; 127: 35.

Lundquist A. Fine needle aspiration biopsy of the liver. *Acta Med Scand* 1971; 520: 1.

Holm H H, Pedersen J F, Kristensen J K, Rasmussen S N, Hancke S, Jensen F. Ultrasonically guided percutaneous puncture. *Radiol Clin North Am* 1975; 13: 493.

von Schreeb T, Arner O, Skovsted G, Wikstad N. Renal adenocarcinoma: Is there a risk of spreading tumor cells in diagnostic puncture? *Scand J Urol Nephrol* 1967; 1: 270.

Berg J W, Robbins G F. A late look at the safety of aspiration biopsy. *Cancer* 1962; 15: 826.

Ferrucci J T Jr, Wittenberg J, Margolies M N, Carey R W. Malignant seeding of the tract after thin-needle aspiration biopsy. *Radiology* 1979; 130: 345.

Smith F P, Macdonald J S, Schein P S, Or-

nitz R D. Cutaneous seeding of pancreatic cancer by skinny-needle aspiration biopsy. *Arch Intern Med* 1980; 140: 855.

Bush W H Jr, Burnett L L, Gibbons R P. Case reports: Needle tract seeding of renal cell carcinoma. *Am J Roentgenol* 1977; 129: 725.

Riska H, Friman C. Fatality after fine-needle aspiration biopsy of liver. (Letter) *Br J Med* 1975; 1: 517.

Evans W K, Ho C S, McLoughlin M J, Tao L C. Fatal necrotizing pancreatitis following fine-needle aspiration biopsy of the pancreas. *Radiology* 1981; 141: 61.

Ferrucci J T Jr, Wittenberg J, Mueller P R, et al. Diagnosis of abdominal malignancy by radiologic fine-needle aspiration. *Am J Roentgenol* 1980; 134: 323.

Schnyder P A, Candardjis G, Anderegg A. Peritonitis after thin needle aspiration biopsy of an abscess. *Am J Roentgenol* 1981; 137: 1271.

Livraghi T, Lombardi C, Mascia G. Bile peritonitis: Another complication after fine-needle biopsy. *Diagn Imaging* 1983; 52: 33.

Schulz T B. Fine-needle biopsy of the liver complicated with bile peritonitis. *Acta Med Scand* 1976; 199: 141.

Otto R. Results of 1000 fine needle punctures guided under real-time sonographic control. *J Belge Radiol* 1982; 65: 193.

Silver T M, Thornbury J R. Pneumothorax: A complication of percutaneous aspiration of upper pole renal masses. *Am J Roentgenol* 1977; 128: 451.

Bigongiari L R. Vagal pseudohemorrhage after percutaneous biopsy. *Invest Radiol* 1980; 15: 350.

Zamcheck N, Klausenstock O. Liver biopsy: The risk of needle biopsy. *N Engl J Med* 1953; 249: 1062.

Kidd R, Freeny P C, Bartha M. Single pass fine-needle aspiration biopsy. *Am J Roentgenol* 1979; 133: 333.

CHAPTER 27

Interventional ultrasound in cancer therapy

H. H. Holm & N. Juul

The ability to place needles precisely in tumors under ultrasonic guidance provides a new possibility of direct percutaneous treatment of malignant lesions in various ways. In principle, a wide variety of different agents can be applied through the needle into the tumor. They include radioactive sources, heat, and a large number of chemical substances including various types of caustics and antineoplastic drugs.

Radioactive sources

Implantation of radioactive seeds in deep-seated lesions has until now required surgical procedures and has not been very precise. To circumvent these drawbacks we have developed a method for percutaneous insertion of radioactive I^{125} seeds into abdominal tumors guided by dynamic ultrasound scanning.

The seeds measure 0.8×4.5 mm and contains Iodine^{125} which has a half-life of 60 days and a half layer value in tissue and lead of 20 mm and 0.025 mm, respectively (Fig. 27-1).

The advantages of "internal" irradiation as opposed to external beam irradiation in deep-seated cancer is the higher total tumor dose which is de-

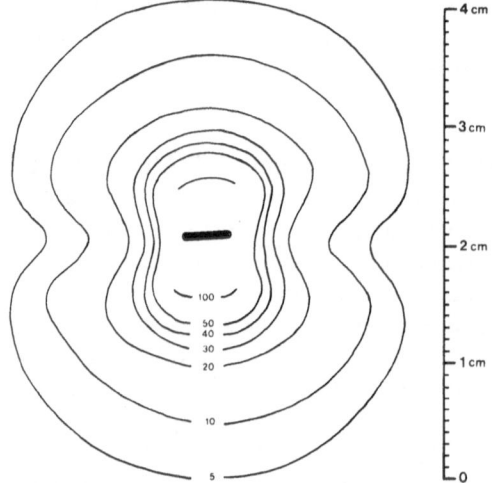

Fig. 27-1. Dose distribution of an I^{125} seed

livered more precisely and at a low dose rate. There is also a lower dose to the normal tissue. Furthermore, the radiation dose is delivered and the treatment completed in a single session, as opposed to multifractionated external beam therapy which may take 6 weeks.

According to Hilaris, seed implantation is the treatment of choice for unresected cancers which are not too widespread, fast growing or highly radiosensitive as to make external X-ray therapy or chemotherapy preferable.

178

Abdominal cancer

A puncture transducer mounted in a special x-y-z coordinate system is used. According to a precalculated treatment program the transducer is brought step-wise (guided by the dynamic ultrasound image) into each of the correct puncture localizations (Fig. 27-2).

The needle containing a predetermined number of I^{125} seeds and cromic catgut spacers is inserted to the deepest part of the tumor (Fig. 27-3). The needle tip echo is generally visible during this procedure. When in place a stopper is swung on top of the stylet and the needle withdrawn.

The localization of the seeds can be visualized in a number of ways. Ultrasonically, radiologically and on scintigraphy, and on this basis dose calculations can be made (Fig. 27-4).

We have treated 13 patients with inoperable malignant abdominal tumors with this technique. A combination of local – and diazepam anesthesia was used and a maximum of 18 needles and 73 seeds were inserted. There were no complications related to the procedure.

One patient had multiple liposarco-

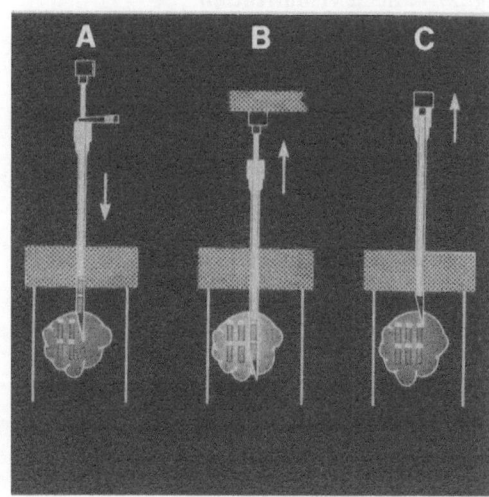

Fig. 27-3. The insertion technique
A: The needle containing a precalculated number of seeds and chromic catgut spacers as well as a stylet is advanced to the deepest area of the tumor. B: A stopper is swung in front of the stylet and the needle is then withdrawn. C: The seeds are thus deposited in the predetermined place.

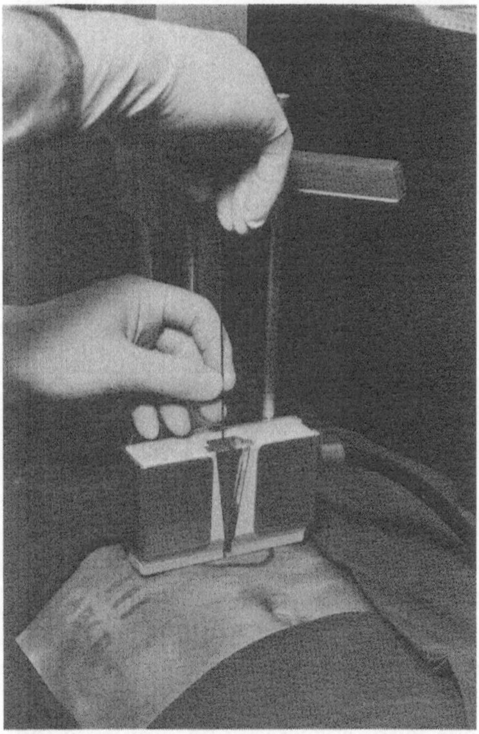

Fig. 27-2. Seed insertion in abdominal cancer
The dynamic scanner is placed in a fixture which allows controlled movements in a x-y-z coordinate system. The scanner is brought step-wise into each puncture location and the insertion performed under direct monitoring of the location of the needle tip.

mas of which 1 was treated with temporary response. Two had large liver metastases which did not respond and 10 patients with pancreatic cancers which fulfilled the selection criteria were treated.

Two of the pancreatic cancers did not

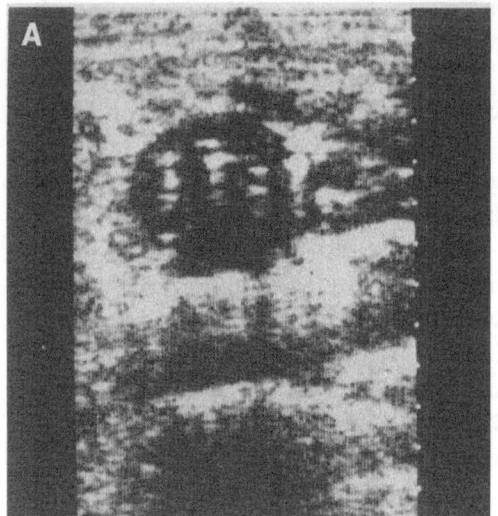

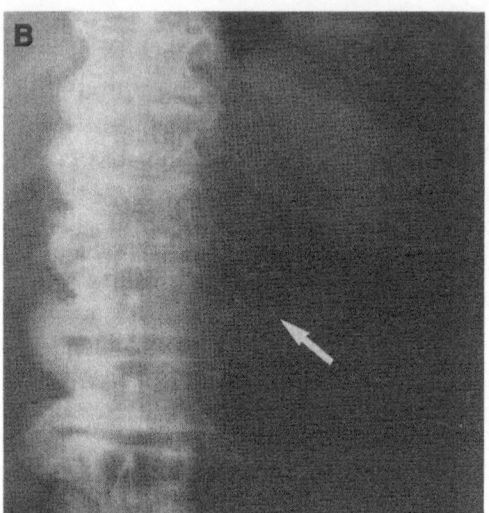

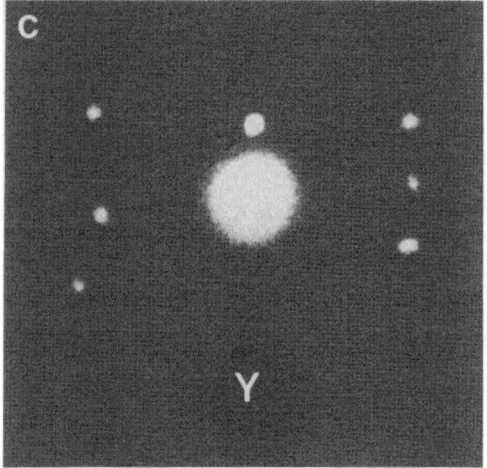

Fig. 27-4. Seed visualization
A: Sonogram showing the seeds inside the tumor. B: Radiogram showing the evenly distributed seeds in the tumor area. C: Scintigram of the implanted seeds demonstrating a rapid fall-off of dose. Lateral body contour and epigastrium marked.

respond at all as judged from the localization of the seeds and the size of the tumor on subsequent plain X-ray and ultrasonic scanning, respectively. Seven tumors showed a response. However, all patients except a patient with a Werner-Morrison syndrome died within 12 months of the implantation.

In 2 of the 7 responding pancreatic cancers no autopsy was performed. In 3 patients residual cancer was found at autopsy.

The most encouraging observation in this series was that in 2 patients who died of unrelated disorders (heart attack and liver abscess) 1 and 7 months later no cancer was detectable at autopsy.

Prostatic cancer

We have developed this technique further and now use it for transperineal im-plantation of I^{125} seeds in low and medium differentiated T-2 prostatic cancers guided by transrectal scanning.

Based upon a series of prostatic scans the number of punctures and seeds are determined and the needles loaded with seeds and spacers. The implantation procedure is carried out in spinal anesthesia with the patient in the lithotomy position. A special puncture attachment with multiple puncture canals is mounted on a B & K rectal scanner which is placed in a fixture (Fig. 27-5).

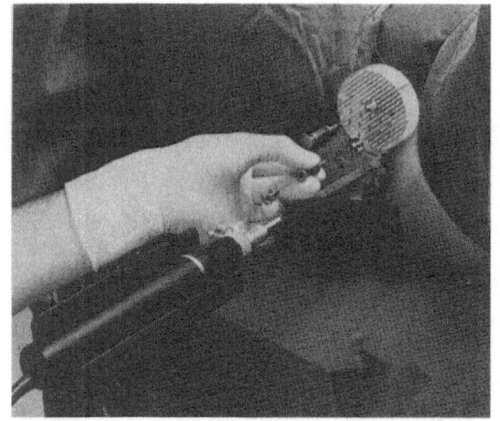

Fig. 27-5. Seed insertion in prostatic cancer
The transrectal scanner with a needle steering device is mounted in a fixture. Under ultrasonic guidance the needles containing the seeds are advanced transperineally into the prostate parallel to the scanner axis.

The insertion technique is shown in Fig. 27-6.

The seeds can be seen on an ultrasound scan of the prostate as well as on a plain X-ray (Fig. 27-7).

The tumor dose delivered by the seeds is 160 Gy, which is very high. The patients are treated postoperatively with an additional external irradiation dose of 40 Gy to the iliac lymph nodes.

Six patients have until now been treated without complications. It was not attempted to avoid puncture of the urethra and 2 patients passed a few seeds during micturition on the 1st day. Five patients have been followed from 1–13 months.

As seen in Fig. 27-8, the prostate vol-

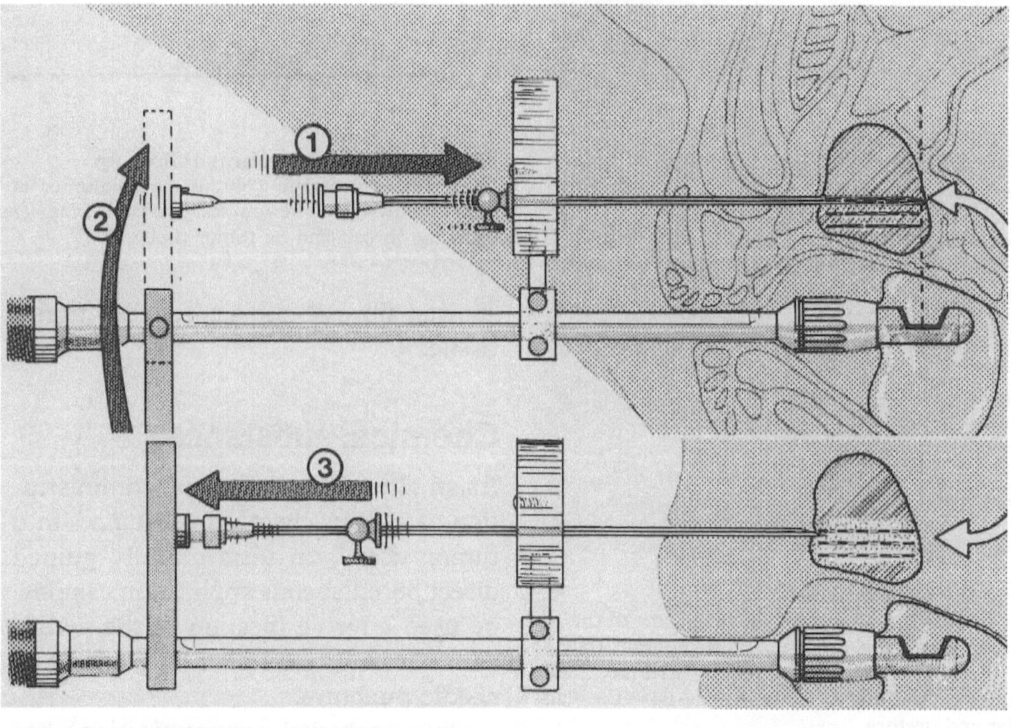

Fig. 27-6. The insertion technique
1: The needle containing seeds, spacers and a stylet is advanced into the scanning plane. This is indicated by a strong echo reflection from the needle tip in the prostatic area. 2: A stopper is swung in front of the stylet. 3: The needle is withdrawn leaving the seeds in the prostate.

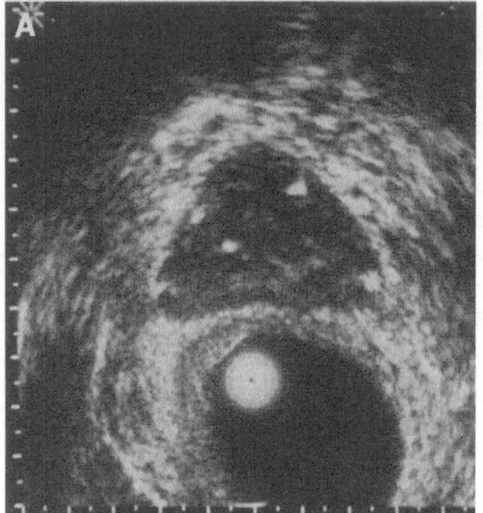

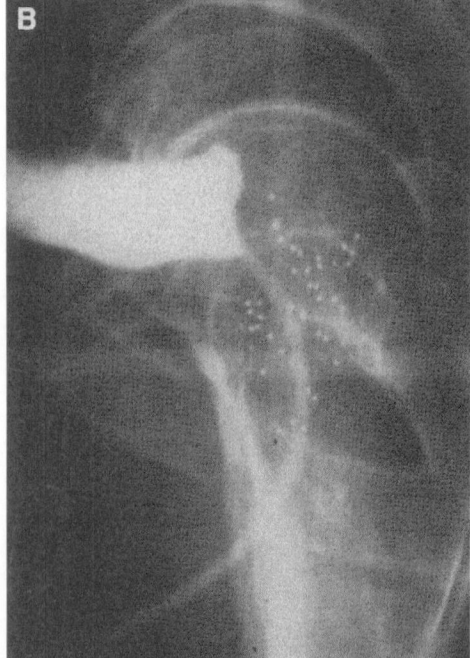

Fig. 27-7. Seed visualization
A: Transrectal sonogram showing some of the I^{125} seeds inside the prostate. B: Radiogram (lateral projection) showing the evenly distributed seeds in the prostatic area. Contrast material in the bladder and urethra.

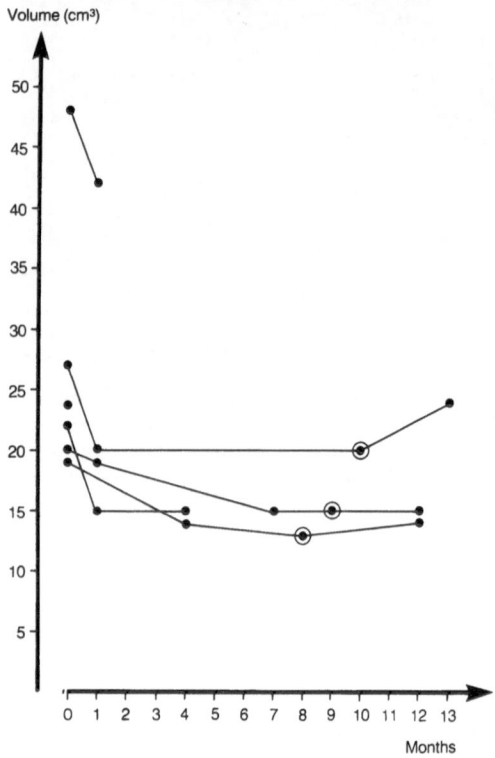

J^{125} seed implantation in prostatic cancer
(6 patients)

Fig. 27-8. Prostatic cancer follow-up
Preoperative prostatic volumes and volumes at each follow-up. The circles indicate biopsies showing fibrosis and no tumor tissue.

ume was reduced in the 5 cases who could be followed. Follow-up biopsies showed fibrosis and no residual tumor tissue.

Chemical substances

As an alternative to arterial administration of various chemical substances into tumor tissue, an ultrasonically guided direct percutaneous application can now be used after verification of the malignant nature of the lesion by guided fine needle puncture.

Ohto (personal communication) has attempted injection of several caustics and antineoplastic drugs directly into liver metastases and now favors concentrated alcohol which results in necrosis and subsequent fibrosis. Guarnieri & Canale have also injected many different chemical substances into liver metastases (heterologous proteins, various caustics, antineoplastic drugs, vasoactive substances, local anesthetics) under ultrasonic guidance. They have repeatedly treated a total of 12 metastases in 4 patients. No complications from the fine needle puncture were observed, while various local and general effects related to the injected compounds were registered.

In collaboration with the onchology department we have injected interferon under ultrasonic guidance in various parts of a recurrent renal cell carcinoma measuring $4 \times 6 \times 8$ cm which did not respond to external irradiation and chemotherapy. The treatment was repeated 3 times and a total of 56 million units interferon were deposited in the mass. The injection caused slight changes in the internal echo pattern of the mass which became slightly more echo-poor. However, no significant effect was noticed with regard to tumor size and the patient died 3 months later of disseminated disease.

Heat

Recently there has been an increasing interest in thermo treatment of malignant tumors either alone or in combination with radiotherapy. However, a fundamental problem is to obtain a sufficiently high temperature inside deeply located tumors without causing damage to the overlying normal structures. To circumvent this problem, invasive microwave antennae which can produce the necessary temperature in-

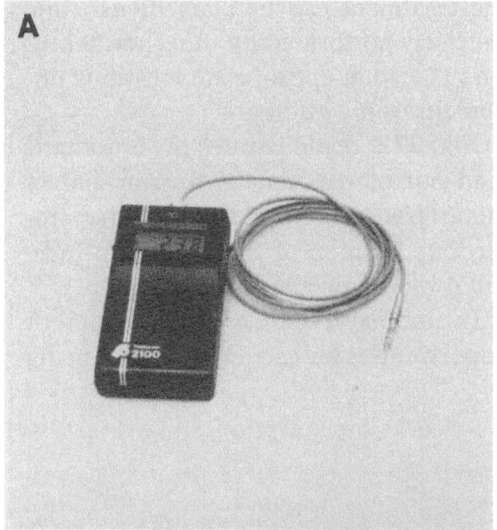

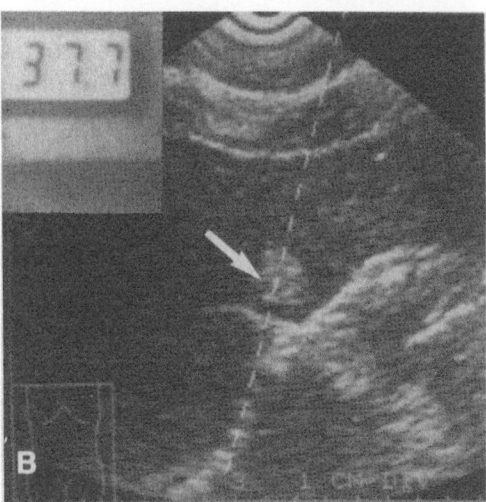

Fig. 27-9. Temperature measurement
A: Digital thermometer with automatic read-out of temperature connected to the stylet of a fine needle. B: The temperature can be read out with the needle tip monitored on the screen. Needle tip indicated.

side the tumor have been developed by Strohbehn and coworkers. Optimal placement of such antennae under ultrasonic guidance is an obvious possibility. Monitoring of the temperature differences inside and outside the tumor during the treatment can be conveniently and precisely performed by thin thermosensors placed at appropriate locations under ultrasonic guidance.

Fig. 27-9 demonstrates the automatic read-out of the temperature in a liver metastasis corresponding to the immediate location of the needle tip indicated by the needle tip echo.

In summary, there are many different opportunities for cancer treatment using interventional ultrasound. One must admit, however, that – maybe with the exception of I^{125} seeds in prostatic cancer – the results obtained until now are far from dramatic. But it should be realized that all known conventional treatments were useless in the patients selected for these new types of treatment. Therefore any marginal gain should be considered a success.

References

Holm H H, Strøyer I, Hansen H, Stadil F. Ultrasonically guided percutaneous interstitial implantation of Iodine 125 seeds in cancer therapy. *Br J Radiol* 1981; 54: 665.

Hilaris B S. Handbook of Interstitial Brachytherapy. Acton, Mass.: Publishing Sciences Group, 1975.

Holm H H, Juul N, Pedersen J F, Hansen H, Strøyer I. Transperineal ^{125}Iodine seed implantation in prostatic cancer guided by transrectal ultrasonography. *J Urol* 1983; 130: 283.

Guarnieri A, Canale M. Ultrasonically guided fine needle drug-infiltration of liver metastases. Third International Congress on Interventional Ultrasound, Copenhagen 1983.

Strohbehn J W, Bowers E D, Douple E B. An invasive microwave antenna for locally-induced hyperthermia for cancer therapy. *J Microwave Power* 1979; 14: 339.

Subject Index

Abscess drainage 156
Amniocentesis 123
– early pregnancy 122
– late pregnancy 129
Aspiration biopsy 21
Aspiration cytology 25
Aspiration handle 19

Bile drainage 56
Bile duct puncture 66

Cancer therapy 178
Cardiac catheterization 164
Chemotherapy, local 182
Cholangiography 54
Cystic lesion 23
Cytology 26

Dynamic scanning 15, 18

Electron microscopy 29

Fatalities 174
Fetal
– blood sampling 122, 124
– heart puncture 127
– hydrocephalus 138
– skin biopsy 122, 124
– therapy 132
– urinary diversion 137
Fluid collections
– diagnostic puncture 155
– therapeutic puncture 156

Gall bladder drainage 61
Gastrointestinal puncture 148
Gynecological puncture 113

Hazards 170
Hemorrhage 172
Hepatic, see also liver
Histological biopsy 19, 22, 35, 37
Hydatid cysts 24

Intracardiac electroplacement 167
Intraoperative puncture 65
Intraoperative scanning 65
Intrauterine transfusion 132, 133

Kidney, see also renal
Kidney, cysts 87
Kidney, solid lesions 84

Liver, see also hepatic
Liver abscess 46
– amebic 47
– pyogenic 46
Liver cysts 44
– simple 45
– hydatic 45
Liver puncture 43
Liver, solid lesions 48

Markerline 17
Microcore biopsy 37
Multistix SG 24

Needles 18
Needle steering devices 17
Needle tract seeding 171, 172
Nephrolithotomy 77
Nephrostomy 72

Oocytes aspiration 117
Ovarian mass lesions 113

Pancreas, cystic lesion 101
Pancreas, solid lesions 101
Pancreatic puncture 100
Pancreatography 106
Pericardial biopsy 164
Pericardial fenestration 164
Pericardiocentesis 160
Prostatic biopsy 94
Puncture guidance 16
Puncture fatalities 171
Puncture precautions 17
Puncture risks 169

Radioactive seeds 178
– – in abdominal cancer 179
– – in prostatic cancer 170
Renal, see also kidney
Renal biopsy 91
Renal cyst 24
Renal puncture 84
Retroperitoneal puncture 143

Safety precautions 20
Scanners 17
Smears
– evalutation 26
– preparation 25
– staining 25
Surecut needle 35

Thermotherapy, local 183
Tumor cell spread 169